T0254922

# Probability and Statistics
## for Data Science
### Math + R + Data

## CHAPMAN & HALL/CRC DATA SCIENCE SERIES

Reflecting the interdisciplinary nature of the field, this book series brings together researchers, practitioners, and instructors from statistics, computer science, machine learning, and analytics. The series will publish cutting-edge research, industry applications, and textbooks in data science.

The inclusion of concrete examples, applications, and methods is highly encouraged. The scope of the series includes titles in the areas of machine learning, pattern recognition, predictive analytics, business analytics, Big Data, visualization, programming, software, learning analytics, data wrangling, interactive graphics, and reproducible research.

Published Titles

**Feature Engineering and Selection: A Practical Approach for Predictive Models**
Max Kuhn and Kjell Johnson

**Probability and Statistics for Data Science: Math + R + Data**
Norman Matloff

# Probability and Statistics
# for Data Science

## Math + R + Data

Norman Matloff

## CRC Press
Taylor & Francis Group
Boca Raton  London  New York

CRC Press is an imprint of the
Taylor & Francis Group, an **informa** business

A CHAPMAN & HALL BOOK

CRC Press
Taylor & Francis Group
6000 Broken Sound Parkway NW, Suite 300
Boca Raton, FL 33487-2742

Printed on acid-free paper

International Standard Book Number-13: 978-1-138-39329-5 (Paperback)
International Standard Book Number-13: 978-0-367-26093-4 (Hardback)

| Library of Congress Cataloging-in-Publication Data |
| --- |
| Names: Matloff, Norman S., author. |
| Title: Probability and statistics for data science / Norman Matloff. |
| Description: Boca Raton : CRC Press, Taylor & Francis Group, 2019. |
| Identifiers: LCCN 2019008196 \| ISBN 9781138393295 (pbk. : alk. paper) |
| Subjects: LCSH: Probabilities--Textbooks. \| Mathematical statistics--Textbooks. \| Probabilities--Data processing. \| Mathematical statistics--Data processing. |
| Classification: LCC QA273 .M38495 2019 \| DDC 519.5--dc23 |
| LC record available at https://lccn.loc.gov/2019008196 |

**Visit the Taylor & Francis Web site at**
**http://www.taylorandfrancis.com**

**and the CRC Press Web site at**
**http://www.crcpress.com**

# Contents

       Functions . . . . . . . . . . . . . . . . . . . . . . . . . . . 115

       6.4.1   CDFs . . . . . . . . . . . . . . . . . . . . . . . . . 115

       6.4.2   Non-Discrete, Non-Continuous Distributions . . . . 119

6.5    Density Functions . . . . . . . . . . . . . . . . . . . . . . 119

       6.5.1   Properties of Densities . . . . . . . . . . . . . . . 120

       6.5.2   Intuitive Meaning of Densities . . . . . . . . . . . 122

       6.5.3   Expected Values . . . . . . . . . . . . . . . . . . . 122

6.6    A First Example . . . . . . . . . . . . . . . . . . . . . . . 123

6.7    Famous Parametric Families of Continuous Distributions . 124

       6.7.1   The Uniform Distributions . . . . . . . . . . . . . 125

               6.7.1.1   Density and Properties . . . . . . . . . . 125

               6.7.1.2   R Functions . . . . . . . . . . . . . . . . 125

               6.7.1.3   Example: Modeling of Disk Performance . 126

               6.7.1.4   Example:  Modeling of Denial-of-Service
                         Attack . . . . . . . . . . . . . . . . . . . 126

       6.7.2   The Normal (Gaussian) Family of Continuous
               Distributions . . . . . . . . . . . . . . . . . . . . 127

               6.7.2.1   Density and Properties . . . . . . . . . . 127

               6.7.2.2   R Functions . . . . . . . . . . . . . . . . 127

               6.7.2.3   Importance in Modeling . . . . . . . . . . 128

       6.7.3   The Exponential Family of Distributions . . . . . . 128

               6.7.3.1   Density and Properties . . . . . . . . . . 128

               6.7.3.2   R Functions . . . . . . . . . . . . . . . . 128

               6.7.3.3   Example: Garage Parking Fees . . . . . . 129

               6.7.3.4   Memoryless Property of Exponential
                         Distributions . . . . . . . . . . . . . . . 130

               6.7.3.5   Importance in Modeling . . . . . . . . . . 131

# III   Multivariate Analysis                                          243

# 11 Multivariate Distributions                                        245

# About the Author

**Dr. Norm Matloff** is a professor of computer science at the University of California at Davis, and was formerly a professor of statistics at that university. He is a former database software developer in Silicon Valley, and has been a statistical consultant for firms such as the Kaiser Permanente Health Plan.

Dr. Matloff was born in Los Angeles, and grew up in East Los Angeles and the San Gabriel Valley. He has a PhD in pure mathematics from UCLA, specializing in probability theory and statistics. He has published numerous papers in computer science and statistics, with current research interests in parallel processing, regression methodology, machine learning, and recommender systems.

Professor Matloff is an award-winning expositor. He is a recipient of the campuswide Distinguished Teaching Award at his university, and his book, *Statistical Regression and Classification: From Linear Models to Machine Learning*, was selected for the international 2017 Ziegel Award. (He also has been a recipient of the campuswide Distinguished Public Service Award at UC Davis.)

# To the Instructor

*Statistics is not a discipline like physics, chemistry or biology where we study a subject to solve problems in the same subject. We study statistics with the main aim of solving problems in other disciplines* — C.R. Rao, one of the pioneers of modern statistics

*The function of education is to teach one to think intensively and to think critically. Intelligence plus character — that is the goal of true education* — Dr. Martin Luther King, American civil rights leader

*[In spite of] innumerable twists and turns, the Yellow River flows east* — Confucius, ancient Chinese philosopher

This text is designed for a junior/senior/graduate-level based course in probability and statistics, *aimed specifically at data science students (including computer science)*. In addition to calculus, the text assumes some knowledge of matrix algebra and rudimentary computer programming.

**But why is this book different from all other books on math probability and statistics?**

Indeed. it *is* quite different from the others. Briefly:

- The subtitle of this book, *Math + R + Data*, immediately signals a difference from other "math stat" books.

- Data Science applications, e.g., random graph models, power law distribution, Hidden Markov models, PCA, Google PageRank, remote sensing, mixture distributions, neural networks, the Curse of Dimensionality, and so on.

- Extensive use of the R language.

The subtitle of this book, *Math + R + Data*, immediately signals that the book follows a very different path. Unlike other "math stat" books, this one has a strong applied emphasis, with lots of real data, facilitated by extensive use of the R language.

The above quotations explain the difference further. First, this book is definitely written from an applications point of view. Second, it pushes the student to think critically about the *how* and *why* of statistics, and to "see the big picture."

- **Use of real data, and early introduction of statistical issues:**

  The Rao quote at the outset of this Preface resonates strongly with me. Though this is a "math stat" book — random variables, density functions, expected values, distribution families, stat estimation and inference, and so on — it takes seriously the Data Science theme claimed in the title, *Probability and Statistics for Data Science*. A book on Data Science, even a mathematical one, should make heavy use of DATA!

  This has implications for the ordering of the chapters. We bring in statistics early, and statistical issues are interspersed throughout the text. Even the introduction to expected value, Chapter 3, includes a simple prediction model, serving as a preview of what will come in Chapter 15. Chapter 5, which covers the famous discrete parametric models, includes an example of fitting the power law distribution to real data. This forms a prelude to Chapter 7, which treats sampling distributions, estimation of mean and variance, bias and so on. Then Chapter 8 covers general point estimation, using MLE and the Method of Moments to fit models to real data. From that point onward, real data is used extensively in every chapter.

  The datasets are all publicly available, so that the instructor can delve further into the data examples.

- **Mathematically correct – yet highly intuitive:**

  The Confucius quote, though made long before the development of formal statistical methods, shows that he had a keen **intuition**, anticipating a fundamental concept in today's world of data science — data smoothing. Development of such strong intuition in our students is a high priority of this book.

  This is of course a mathematics book. All models, concepts and so on are described precisely in terms of random variables and distributions. In addition to calculus, matrix algebra plays an important role. Optional Mathematical Complements sections at the ends of many chapters allow inquisitive readers to explore more sophisticated material. The mathematical exercises range from routine to more challenging.

On the other hand, this book is not about "math for math's sake." In spite of being mathematically precise in description, it is definitely not a theory book.

For instance, the book does not define probability in terms of sample spaces and set-theoretic terminology. In my experience, defining probability in the classical manner is a major impediment to learning the intuition underlying the concepts, and later to doing good applied work. Instead, I use the intuitive, informal approach of defining probability in terms of long-run frequency, in essence taking the Strong Law of Large Numbers as an axiom.

I believe this approach is especially helpful when explaining conditional probability and expectation, concepts that students notoriously have trouble with. Under the classical approach, students have trouble recognizing when an exercise — and more importantly, an actual application — calls for a conditional probability or expectation if the wording lacks the explicit phrase *given that*. Instead, I have the reader think in terms of repeated trials, "How often does A occur *among those times* in which B occurs?", which is easier to relate to practical settings.

- **Empowering students for real-world applications:**

The word *applied* can mean different things to different people. Consider for instance the interesting, elegant book for computer science students by Mitzenmacher and Upfal [33]. It focuses on probability, in fact discrete probability, and its intended class of applications is actually the *theory* of computer science.

I instead focus on the actual use of the material in the real world; which tends to be more continuous than discrete, and more in the realm of statistics than probability. This is especially valuable, as Big Data and Machine Learning now play a significant role in computer and data science.

One sees this philosophy in the book immediately. Instead of starting out with examples involving dice or coins, the book's very first examples involve a model of a bus transportation system and a model of a computer network. There are indeed also examples using dice, coins and games, but the theme of the late Leo Breiman's book subtitle [5], "With a View toward Applications," is never far away.

If I may take the liberty of extending King's quote, I would note that today statistics is a core intellectual field, affecting virtually everyone's daily lives. The ability to use, or at the very least *understand*, statistics is vital to good citizenship, and as an author I take this as a mission.

- **Use of the R Programming Language:**

  The book makes use of some light programming in R, for the purposes of simulation and data analysis. The student is expected to have had some rudimentary prior background in programming, say in one of Python, C, Java or R, but no prior experience with R is assumed. A brief introduction is given in the book's appendix, and some further R topics are interspered with the text as Computational Complements.

  R is widely used in the world of statistics and data science, with outstanding graphics/visualization capabilities, and a treasure chest of more than 10,000 contributed code packages.

  Readers who happen to be in computer science will find R to be of independent interest from a CS perspective. First, R follows the *functional language* and *object-oriented* paradigms: Every action is implemented as a function (even '+'); side effects are (almost) always avoided; functions are first-class objects; several different kinds of class structures are offered. R also offers various interesting metaprogramming capabilities. In terms of programming support, there is the extremely popular RStudio IDE, and for the "hard core" coder, the Emacs Speaks Statistics framework. Most chapters in the book have Computational Complements sections, as well as a Computational and Data Problems portion in the exercises.

**Chapter Outline:**

Part I, Chapters 1 through 6: These introduce probability, Monte Carlo simulation, discrete random variables, expected value and variance, and parametric families of discrete distributions.

Part II, Chapters 7 through 10: These then introduce statistics, such as sampling distributions, MLE, bias, Kolmogorov-Smirnov and so on, illustrated by fitting gamma and beta density models to real data. Histograms are viewed as density estimators, and kernel density estimation is briefly covered. This is followed by material on confidence intervals and significance testing.

Part III, Chapters 11 through 17: These cover multivariate analysis in various aspects, such as multivariate distribution, mixture distributions,

PCA/log-linear model, dimension reduction, overfitting and predictive analytics. Again, real data plays a major role.

**Coverage Strategies:**

The book can be comfortably covered in one semester. If a more leisurely pace is desired, or one is teaching under a quarter system, the material has been designed so that some parts can be skipped without loss of continuity. In particular, a more statistics-oriented course might omit the material on Markov chains, while a course focusing more on machine learning may wish to retain this material (e.g., for Hidden Markov models). Individual sections on specialty topics also have been written so as not to create obstacles later on if they are skipped.

Chapter 11 on multivariate distributions is very useful for data science, e.g., for its relation to clustering. However, instructors who are short on time or whose classes may not have a strong background in matrix algebra may safely skip much of this material.

**A Note on Typography**

In order to help the reader keep track of the various named items, I use math italics for mathematical symbols and expressions, and bold face for program variable and function names. I include R package names for the latter, except for those beginning with a capital letter.

**Thanks:**

The following, among many, provided valuable feedback for which I am very grateful: Ibrahim Ahmed; Ahmed Ahmedin; Stuart Ambler; Earl Barr; Benjamin Beasley; Matthew Butner; Vishal Chakraborti, Michael Clifford; Dipak Ghosal; Noah Gift; Laura Matloff; Nelson Max, Deep Mukhopadhyay, Connie Nguyen, Jack Norman, Richard Oehrle, Michael Rea, Sana Vaziri, Yingkang Xie, and Ivana Zetko. My editor, John Kimmel, is always profoundly helpful. And as always, my books are also inspired tremendously by my wife Gamis and daughter Laura.

# To the Reader

*I took a course in speed reading, and read War and Peace in 20 minutes. It's about Russia* — comedian Woody Allen

*I learned very early the difference between knowing the name of something and knowing something* — Richard Feynman, Nobel laureate in physics

*Give me six hours to chop down a tree and I will spend the first four sharpening the axe* — Abraham Lincoln

This is NOT your ordinary math or programming book.

In order to use this material in real-world applications, it's crucial to understand what the math *means*, and what the code actually *does*.

In this book, you will often find several consecutive paragraphs, maybe even a full page, in which there is no math, no code and no graphs. Don't skip over these portions of the book! They may actually be the most important ones in the book, in terms of your ability to apply the material in the real world.

And going hand-in-hand with this point, mathematical intuition is key. As you read, stop and think about the intuition underlying those equations.

A closely related point is that the math and code complement each other. Each will give you deeper insight in the other. It may at first seem odd that the book intersperses math and code, but sooon you will find their interaction to be quite helpful to your understanding of the material.

**The "Plot"**

Think of this book as a movie. In order for the "plot" to work well, we will need preparation. This book is aimed at applications to Data Science, so the ultimate destination of the "plot" is statistics and predictive analytics.

The foundation for those fields is probability, so we lay the foundation first in Chapters 1 through 6. We'll need more probability later — Chapters 9 and 11 — but in order to bring in some "juicy" material into the "movie" as early as possible, we introduce statistics, especially analysis of real DATA, in Chapters 7 and 8 at this early stage.

The final chapter, on Markov chains, is like a "sequel" to the movie. This sets up some exciting Data Science applications such as Hidden Markov Models and Google's PageRank search engine.

# Part I

# Fundamentals of Probability

# Chapter 1

# Basic Probability Models

This chapter will introduce the general notions of probability. Most of it will seem intuitive to you, and intuition is indeed crucial in the field of probability and statistics. On the other hand, do not rely on intuition alone; pay careful attention to the general principles which are developed. In more complex settings intuition may not be enough, or may even mislead you. The tools discussed here will be essential, and **will be cited frequently throughout the book**.

In this book, we will be discussing both "classical" probability examples involving coins, cards and dice, and also examples involving applications in the real world. The latter will involve diverse fields such as data mining, machine learning, computer networks, bioinformatics, document classification, medical fields and so on. Applied problems actually require a bit more work to fully absorb, but needless to say, you will derive the most benefit from those examples rather than ones involving coins, cards and dice.[1]

Let's start with one concerning transportation.

## 1.1   Example: Bus Ridership

Consider the following analysis of bus ridership, which (in more complex form) could be used by the bus company/agency to plan the number of buses, frequency of stops and so on. Again, in order to keep things easy, it

---

[1]Well, what about gambling? Isn't that an "application"? Yes, but those are actually some of the deepest, most difficult applications.

will be quite oversimplified, but the principles will be clear.

Here is the model:

- At each stop, each passsenger alights from the bus, independently of the actions of others, with probability 0.2 each.

- Either 0, 1 or 2 new passengers get on the bus, with probabilities 0.5, 0.4 and 0.1, respectively. Passengers at successive stops act independently.

- Assume the bus is so large that it never becomes full, so the new passengers can always board.

- Suppose the bus is empty when it arrives at its first stop.

Here and throughout the book, **it will be greatly helpful to first name the quantities or events involved**. Let $L_i$ denote the number of passengers on the bus as it *leaves* its $i^{th}$ stop, $i = 1, 2, 3, ...$ Let $B_i$ denote the number of new passengers who board the bus at the $i^{th}$ stop.

We will be interested in various probabilities, such as the probability that no passengers board the bus at the first three stops, i.e.,

$$P(B_1 = B_2 = B_3 = 0)$$

The reader may correctly guess that the answer is $0.5^3 = 0.125$. But again, we need to do this properly. In order to make such calculations, we must first set up some machinery, in the next section.

Again, note that this is a very simple model. For instance, we are not taking into account day of the week, month of the year, weather conditions, etc.

## 1.2   A "Notebook" View: the Notion of a Repeatable Experiment

It's crucial to understand what that 0.125 figure really means in an intuitive sense. To this end, let's put the bus ridership example aside for a moment, and consider the "experiment" consisting of rolling two dice, say a blue one and a yellow one. Let $X$ and $Y$ denote the number of dots we get on the blue and yellow dice, respectively, and consider the meaning of $P(X + Y = 6) = \frac{5}{36}$.

Table 1.1: Sample Space for the Dice Example

| 1,1 | 1,2 | 1,3 | 1,4 | 1,5 | 1,6 |
|-----|-----|-----|-----|-----|-----|
| 2,1 | 2,2 | 2,3 | 2,4 | 2,5 | 2,6 |
| 3,1 | 3,2 | 3,3 | 3,4 | 3,5 | 3,6 |
| 4,1 | 4,2 | 4,3 | 4,4 | 4,5 | 4,6 |
| 5,1 | 5,2 | 5,3 | 5,4 | 5,5 | 5,6 |
| 6,1 | 6,2 | 6,3 | 6,4 | 6,5 | 6,6 |

## 1.2.1 Theoretical Approaches

In the mathematical theory of probability, we talk of a *sample space*, which (in simple cases) consists of a list of the possible outcomes $(X, Y)$, seen in Table 1.1. In a theoretical treatment, we place weights of 1/36 on each of the points in the space, reflecting the fact that each of the 36 points is equally likely, and then say, "What we mean by $P(X + Y = 6) = \frac{5}{36}$ is that the outcomes (1,5), (2,4), (3,3), (4,2), (5,1) have total weight 5/36."

Unfortunately, the notion of sample space becomes mathematically tricky when developed for more complex probability models. Indeed, it requires graduate-level math, called *measure theory*.

And much worse, under the sample space approach, **one loses all the intuition**. In particular, **there is no good way using set theory to convey the intuition underlying conditional probability** (to be introduced in Section 1.3). The same is true for expected value, a central topic to be introduced in Section 3.5.

In any case, most probability computations do not rely on explicitly writing down a sample space. In this particular example, involving dice, it is useful for us as a vehicle for explaining the concepts, but we will NOT use it much.

## 1.2.2 A More Intuitive Approach

But the intuitive notion—which is FAR more important—of what $P(X + Y = 6) = \frac{5}{36}$ means is the following. Imagine doing the experiment many, many times, recording the results in a large notebook:

Table 1.2: Notebook for the Dice Problem

| notebook line | outcome | blue+yellow = 6? |
|---|---|---|
| 1 | blue 2, yellow 6 | No |
| 2 | blue 3, yellow 1 | No |
| 3 | blue 1, yellow 1 | No |
| 4 | blue 4, yellow 2 | Yes |
| 5 | blue 1, yellow 1 | No |
| 6 | blue 3, yellow 4 | No |
| 7 | blue 5, yellow 1 | Yes |
| 8 | blue 3, yellow 6 | No |
| 9 | blue 2, yellow 5 | No |

- Roll the dice the first time, and write the outcome on the first line of the notebook.

- Roll the dice the second time, and write the outcome on the second line of the notebook.

- Roll the dice the third time, and write the outcome on the third line of the notebook.

- Roll the dice the fourth time, and write the outcome on the fourth line of the notebook.

- Imagine you keep doing this, thousands of times, filling thousands of lines in the notebook.

The first 9 lines of the notebook might look like Table 1.2. Here 2/9 (or 8/36) of these lines say Yes. But after many, many repetitions, approximately 5/36 of the lines will say Yes. For example, after doing the experiment 720 times, approximately $\frac{5}{36} \times 720 = 100$ lines will say Yes.

This is what probability really is: In what fraction of the lines does the event of interest happen? **It sounds simple, but if you always think about this "lines in the notebook" idea, probability problems are a lot easier to solve.** And it is the fundamental basis of computer simulation.

Many raeeders will have had a bit of prior exposure to probability principles, and thus be puzzled by the lack of *Venn diagrams* (pictures involving set intersection and union) in this book. Such readers are asked to be patient; they should soon find the approach here to be clearer and more powerful. By the way, it won't be much of an issue after this chapter anyway, since either approach would prepare the student for the coming material.

## 1.3 Our Definitions

If we were to ask any stranger on the street, "What do we mean when we say that the probability of winning some casino game is 20%", she would probably say, "Well, if we were to play the game repeatably, we'd win 20% of the time." **This is actually the way we will define probability in this book**.

The definitions here are intuitive, rather than rigorous math, but intuition is what we need. Keep in mind that we are making <u>definitions</u> below, not a listing of properties.

- We assume an "experiment" which is (at least in concept) repeatable. The above experiment of rolling two dice is repeatable, and even the bus ridership model is so: Each day's ridership record would be a repetition of the experiment.

  On the other hand, the econometricians, in forecasting 2009, cannot "repeat" 2008. Yet all of the econometricians' tools assume that events in 2008 were affected by various sorts of randomness, and we think of repeating the experiment in a conceptual sense.

- We imagine performing the experiment a large number of times, recording the result of each repetition on a separate line in a notebook.

- We say $A$ is an *event* for this experiment if it is a possible boolean (i.e., yes-or-no) outcome of the experiment. In the above example, here are some events:

  * $X + Y = 6$
  * $X = 1$
  * $Y = 3$
  * $X - Y = 4$

- A *random variable* is a numerical outcome of the experiment, such as $X$ and $Y$ here, as well as $X + Y$, $2XY$ and even $\sin(XY)$.

- For any event of interest $A$, imagine a column on $A$ in the notebook. The $k^{th}$ line in the notebook, $k = 1, 2, 3, ...$, will say Yes or No, depending on whether $A$ occurred or not during the $k^{th}$ repetition of the experiment. For instance, we have such a column in our table above, for the event {blue+yellow $= 6$}.

- For any event of interest $A$, we define $P(A)$ to be the long-run fraction of lines with Yes entries.

- For any events $A$ and $B$, imagine a new column in our notebook, labeled "A and B." In each line, this column will say Yes if and only if there are Yes entries for both $A$ and $B$.

  $P(A$ and $B)$ is then defined to be the long-run fraction of lines with Yes entries in the new column labeled "A and B."[2]

- For any events $A$ and $B$, imagine a new column in our notebook, labeled "A or B." In each line, this column will say Yes if and only if at least one of the entries for $A$ and $B$ says Yes.[3]

  $P(A$ or $B)$ is then defined to be the long-run fraction of lines with Yes entries in the new column labeled "A or B."

- For any events $A$ and $B$, imagine a new column in our notebook, labeled "A | B" and pronounced "A given B." In each line:

  * This new column will say "NA" ("not applicable") if the B entry is No.

  * If it is a line in which the B column says Yes, then this new column will say Yes or No, depending on whether the A column says Yes or No.

  Then $P(A|B)$ means the long-run fraction of lines in the notebook in which the A | B column says Yes—**among the lines which do NOT say NA.**

**A hugely common mistake is to confuse** $P(A$ and $B)$ **and** $P(A|B)$. This is where the notebook view becomes so important. Compare the quantities $P(X = 1$ and $S = 6) = 1/36$ and $P(X = 1|S = 6) = 1/5$, where S = X+Y:[4]

---

[2]In most textbooks, what we call "A and B" here is written $A \cap B$, indicating the intersection of two sets in the sample space. But again, we do not take a sample space point of view here.

[3]In the sample space approach, this is written $A \cup B$.

[4]Think of adding an S column to the notebook too

- After a large number of repetitions of the experiment, approximately 1/36 of the lines of the notebook will have the property that both X = 1 and S = 6 (since X = 1 and S = 6 is equivalent to X = 1 and Y = 5).

- After a large number of repetitions of the experiment, if **we look only at the lines in which S = 6**, then **among those lines**, approximately 1/5 of **those lines** will show X = 1.

The quantity P(A|B) is called the *conditional probability of A, given B*.

Note that *and* has higher logical precedence than *or*, e.g., $P(A$ and $B$ or $C)$ means $P[(A$ and $B)$ or $C]$.

Here are some more very important definitions and properties:

- **Definition 1** *Suppose A and B are events such that it is impossible for them to occur in the same line of the notebook. They are said to be* **disjoint** *events.*

- If $A$ and $B$ are disjoint events, then[5]

$$P(A \text{ or } B) = P(A) + P(B) \tag{1.1}$$

By writing

$$\{A \text{ or } B \text{ or } C\} = \{A \text{ or } [B \text{ or } C]\} = \tag{1.2}$$

(1.1) can be iterated, e.g.,

$$P(A \text{ or } B \text{ or } C) = P(A) + P(B) + P(C) \tag{1.3}$$

and

$$P(A_1 \text{ or } A_2 \ldots \text{ or } A_k) = \sum_{i=1}^{k} P(A_k) \tag{1.4}$$

if the events $A_i$ are disjoint.

---

[5]Again, this terminology *disjoint* stems from the set-theoretic sample space approach, where it means that $A \cap B = \phi$. That mathematical terminology works fine for our dice example, but in my experience people have major difficulty applying it correctly in more complicated problems. This is another illustration of why I put so much emphasis on the "notebook" framework.

- If $A$ and $B$ are not disjoint, then

$$P(A \text{ or } B) = P(A) + P(B) - P(A \text{ and } B) \qquad (1.5)$$

In the disjoint case, that subtracted term is 0, so (1.5) reduces to (1.1).

Unfortunately, there is no nice, compact form similar to (1.4) that generalizes (1.5) to $k$ non-disjoint events.

- **Definition 2** *Events $A$ and $B$ are said to be stochastically independent, usually just stated as independent,[6] if*

$$P(A \text{ and } B) = P(A) \cdot P(B) \qquad (1.6)$$

And in general,

$$P(A_1 \text{ and } A_2 \text{ ... and } A_k) = \prod_{i=1}^{k} P(A_k) \qquad (1.7)$$

if the events $A_i$ are independent.

In calculating an "and" probability, how does one know whether the events are independent? The answer is that this will typically be clear from the problem. If we toss the blue and yellow dice, for instance, it is clear that one die has no impact on the other, so events involving the blue die are independent of events involving the yellow die.

On the other hand, in the bus ridership example, it's clear that events involving, say, $L_5$ are NOT independent of those involving $L_6$. For instance, if $L_5 = 0$, it is impossible for $L_6$ to be 3, since in this simple model at most 2 passengers can board at any stop.

- If A and B are not independent, the equation (1.6) generalizes to

$$P(A \text{ and } B) = P(A)P(B|A) \qquad (1.8)$$

This should make sense to you. Suppose 30% of all UC Davis students are in engineering, and 20% of all engineering majors are female. That would imply that 0.30 x 0.20 = 0.06, i.e., 6% of all UCD students are female engineers.

Note that if $A$ and $B$ actually are independent, then $P(B|A) = P(B)$, and (1.8) reduces to (1.6).

---

[6]The term *stochastic* is just a fancy synonym for *random*.

Note too that (1.8) implies

$$P(B|A) = \frac{P(A \text{ and } B)}{P(A)} \qquad (1.9)$$

## 1.4 "Mailing Tubes"

*If I ever need to buy some mailing tubes, I can come here*—friend of the author's, while browsing through an office supplies store

Examples of the above properties, e.g., (1.6) and (1.8), will be given starting in Section 1.6.1. But first, a crucial strategic point in learning probability must be addressed.

Some years ago, a friend of mine was in an office supplies store, and he noticed a rack of mailing tubes. My friend made the remark shown above. Well, (1.6) and 1.8 are "mailing tubes" — make a mental note to yourself saying, "If I ever need to find a probability involving *and*, one thing I can try is (1.6) and (1.8)." **Be ready for this!**

This mailing tube metaphor will be mentioned often.

## 1.5 Example: Bus Ridership Model (cont'd.)

Armed with the tools in the last section, let's find some probabilities. First, let's formally calculate the probability that no passengers board the bus at the first three stops. That's easy, using (1.7). Remember, the probability that 0 passengers board at a stop is 0.5.

$$P(B_1 = 0 \text{ and } B_2 = 0 \text{ and } B_3 = 0) = 0.5^3 \qquad (1.10)$$

Now let's compute the probability that the bus leaves the second stop empty. Again, **we must translate this to math first**, i.e., $P(L_2 = 0)$.

To calculate this, we employ a very common approach:

- Ask, "How can this event happen?"

- Break big events into small events.

- Apply the mailing tubes.

What are the various ways that $L_2$ can be 0? Write the event $L_2 = 0$ as

$$\underbrace{B_1 = 0 \text{ and } L_2 = 0}_{} \text{ or } \underbrace{B_1 = 1 \text{ and } L_2 = 0}_{} \text{ or } \underbrace{B_1 = 2 \text{ and } L_2 = 0}_{} \quad (1.11)$$

The underbraces here do not represent some esoteric mathematical operation. They are there simply to make the grouping clearer — we have two **ors**, so we can use (1.4).

$$P(L_2 = 0) = \sum_{i=0}^{2} P(B_1 = i \text{ and } L_2 = 0) \quad (1.12)$$

And now use (1.8)

$$
\begin{aligned}
P(L_2 = 0) &= \sum_{i=0}^{2} P(B_1 = i \text{ and } L_2 = 0) & (1.13) \\
&= \sum_{i=0}^{2} P(B_1 = i) \, P(L_2 = 0 | B_1 = i) & (1.14) \\
&= 0.5^2 + (0.4)(0.2)(0.5) + (0.1)(0.2^2)(0.5) & (1.15) \\
&= 0.292 & (1.16)
\end{aligned}
$$

For instance, where did that first term, $0.5^2$, come from? Well, $P(B_1 = 0) = 0.5$, and what about $P(L_2 = 0 | B_1 = 0)$? If $B_1 = 0$, then the bus approaches the second stop empty. For it to then *leave* that second stop empty, it must be the case that $B_2 = 0$, which has probability 0.5. In other words, $P(L_2 = 0 | B_1 = 0) = 0.5$.

What about the second term? First, $P(B_1 = 1) = 0.4$. Next, to evaluate $P(L_2 = 0 \mid B_1 = 1)$, reason as follows: If $B_1$ is 1, then the bus *arrives* at the second stop with one passenger. In order to then *leave* that stop with no passengers, it must be that the arriving passenger alights from the bus (probability 0.2) and no new passengers board (probability 0.5). Hence $P(L_2 = 0 \mid B_1 = 1) = (0.2)(0.4)$.

As another example, suppose we are told that the bus arrives empty at the third stop. What is the probability that exactly two people boarded the bus at the first stop?

Note first what is being asked for here: $P(B_1 = 2 | L_2 = 0)$ — a *conditional* probability! Then we have, using (1.9) and (1.8),

$$
\begin{aligned}
P(B_1 = 2 \mid L_2 = 0) &= \frac{P(B_1 = 2 \text{ and } L_2 = 0)}{P(L_2 = 0)} \\
&= \frac{P(B_1 = 2) \, P(L_2 = 0 \mid B_1 = 2)}{0.292} \quad (1.17) \\
&= 0.1 \cdot 0.2^2 \cdot 0.5 / 0.292 \quad (1.18)
\end{aligned}
$$

(the 0.292 had been previously calculated in (1.16)).

Now let's find the probability that fewer people board at the second stop than at the first. Again, first translate that to math, $P(B_2 < B_1)$, and "break big events into small events: The event $B_2 < B_1$ can be written as

$$
\underbrace{B_1 = 1 \text{ and } B_2 < B_1}_{} \text{ or } \underbrace{B_1 = 2 \text{ and } B_2 < B_1}_{} \quad (1.19)
$$

Then follow the same pattern as above to obtain

$$
P(B_2 < B_1) = 0.4 \cdot 0.5 + 0.1 \cdot (0.5 + 0.4) \quad (1.20)
$$

How about this one? Someone tells you that as she got off the bus at the second stop, she saw that the bus then left that stop empty. Let's find the probability that she was the only passenger when the bus left the first stop:

We are given that $L_2 = 0$. But we are *also* given that $L_1 > 0$. Then

$$
\begin{aligned}
P(L_1 = 1 | L_2 = 0 \text{ and } L_1 > 0) &= \frac{P(L_1 = 1 \text{ and } L_2 = 0 \text{ and } L_1 > 0)}{P(L_2 = 0 \text{ and } L_1 > 0)} \\
&= \frac{P(L_1 = 1 \text{ and } L_2 = 0)}{P(L_2 = 0 \text{ and } L_1 > 0)} \quad (1.21)
\end{aligned}
$$

Let's first consider how to get the numerator from the preceding equation. Ask the usual question: How can it happen? In this case, how can the event

$$
L_1 = 1 \text{ and } L_2 = 0 \quad (1.22)
$$

occur? Since we know a lot about the probabilistic behavior of the $B_i$, let's try to recast that event: the event is equivalent to the event

$$B_1 = 1 \text{ and } L_2 = 0 \tag{1.23}$$

which we found before to have probability $(0.4)(0.2)(0.5)$.

It remains to calculate the denominator in (1.21). Here we recast the event

$$L_2 = 0 \text{ and } L_1 > 0 \tag{1.24}$$

in terms of the $B_i$? Well, $L_1 > 0$ means that $B_1$ is either 1 or 2. Then break things down in a manner similar to previous computations.

## 1.6   Example: ALOHA Network

In this section, an example from computer networks is presented which, as with the bus ridership example, will be used at a number of points in this book. Probability analysis is used extensively in the development of new, faster types of networks.

We speak of *nodes* on a network. These might be computers, printers or other equipment. We will also speak of *messages*; for simplicity, let's say a message consists of a single character. If a user at a computer hits the N key, say, in a connection with another computer, the user's machine sends the ASCII code for that character onto the network. Of course, this was all transparent to the user, actions behind the scenes.

Today's Ethernet evolved from an experimental network developed at the University of Hawaii, called ALOHA. A number of network nodes would occasionally try to use the same radio channel to communicate with a central computer. The nodes couldn't hear each other, due to the obstruction of mountains between them. If only one of them made an attempt to send, it would be successful, and it would receive an acknowledgement message in response from the central computer. But if more than one node were to transmit, a *collision* would occur, garbling all the messages. The sending nodes would timeout after waiting for an acknowledgement that never came, and try sending again later. To avoid having too many collisions, nodes would engage in random *backoff*, meaning that they would refrain from sending for a while even though they had something to send.

One variation is *slotted* ALOHA, which divides time into intervals which I will call "epochs." Each epoch will have duration 1.0, so epoch 1 consists of the time interval [0.0,1.0), epoch 2 is [1.0,2.0) and so on.

In the simple model we will consider here, in each epoch, if a node is *active*, i.e., has a message to send, it will either send or refrain from sending, with probability $p$ and $1-p$. The value of $p$ is set by the designer of the network. (Real Ethernet hardware does something like this, using a random number generator inside the chip.) Note that a small value of $p$ will tend to produce longer backoff times. The designer may thus choose a small value of $p$ if heavy traffic is anticipated on the network.

The other parameter $q$ in our model is the probability that a node which had been inactive generates a message during an epoch, i.e., the probability that the user hits a key, and thus becomes "active." Think of what happens when you are at a computer. You are not typing constantly, and when you are not typing, the time until you hit a key again will be random. Our parameter $q$ models that randomness; the heavier the network traffic, the large the value of $q$.

Let $n$ be the number of nodes, which we'll assume for simplicity is 2. Assume also for simplicity that the timing is as follows:

- A new message at a node is generated only in the middle of an epoch, say time 8.5.

- The node's decision as to whether to send versus back off is made near the end of an epoch, 90% into the epoch, e.g., time 3.9..

*Example:* Say at the beginning of the epoch which extends from time 15.0 to 16.0, node A has something to send but node B does not. At time 15.5, node B will either generate a message to send or not, with probability $q$ and $1 - q$, respectively. Suppose B does generate a new message. At time 15.9, node A will either try to send or refrain, with probability $p$ and $1 - p$, and node B will do the same. Suppose A refrains but B sends. Then B's transmission will be successful, and at the start of epoch 16 B will be inactive, while node A will still be active. On the other hand, suppose both A and B try to send at time 15.9; both will fail, and thus both will be active at time 16.0, and so on.

Be sure to keep in mind that in our simple model here, during the time a node is active, it won't generate any additional new messages.

Let's observe the network for two epochs, epoch 1 and epoch 2. Assume that the network consists of just two nodes, called node 1 and node 2, both of which start out active. Let $X_1$ and $X_2$ denote the numbers of active nodes at the *very end* of epochs 1 and 2, *after possible transmissions*. We'll take $p$ to be 0.4 and q to be 0.8 in this example.

Please keep in mind that the notebook idea is simply a vehicle to help you understand what the concepts really mean. This is crucial for your intuition and your ability to apply this material in the real world. But the notebook idea is NOT for the purpose of calculating probabilities. Instead, we use the properties of probability, as seen in the following.

## 1.6.1   ALOHA Network Model Summary

- We have $n$ network nodes, sharing a common communications channel.

- Time is divided in epochs. $X_k$ denotes the number of active nodes at the end of epoch $k$, which we will sometimes refer to as the *state* of the system in epoch $k$.

- If two or more nodes try to send in an epoch, they collide, and the message doesn't get through.

- We say a node is active if it has a message to send.

- If a node is active node near the end of an epoch, it tries to send with probability $p$.

- If a node is inactive at the beginning of an epoch, then at the middle of the epoch it will generate a message to send with probability $q$.

- In our examples here, we have $n = 2$ and $X_0 = 2$, i.e., both nodes start out active.

## 1.6.2   ALOHA Network Computations

Let's find $P(X_1 = 2)$, the probability that $X_1 = 2$, and then get to the main point, which is to ask what we really mean by this probability.

How could $X_1 = 2$ occur? There are two possibilities:

- both nodes try to send; this has probability $p^2$

- neither node tries to send; this has probability $(1-p)^2$

Thus

$$P(X_1 = 2) = p^2 + (1-p)^2 = 0.52 \qquad (1.25)$$

Let's look at the details, using our definitions. Once again, **it is helpful to name some things.** Let $C_i$ denote the event that node i tries to send, i = 1,2. Then using the definitions in Section 1.3, our steps would be

$$
\begin{aligned}
P(X_1 = 2) &= P(\underbrace{C_1 \text{ and } C_2}_{} \text{ or } \underbrace{\text{not } C_1 \text{ and not } C_2}_{}) & (1.26) \\
&= P(C_1 \text{ and } C_2) + P(\text{ not } C_1 \text{ and not } C_2) & (1.27) \\
&= P(C_1)P(C_2) + P(\text{ not } C_1)P(\text{ not } C_2) & (1.28) \\
&= p^2 + (1-p)^2 & (1.29)
\end{aligned}
$$

Here are the reasons for these steps:

(1.26): We listed the ways in which the event $\{X_1 = 2\}$ could occur.

(1.27): Write

$$G = C_1 \text{ and } C_2$$

and

$$H = D_1 \text{ and } D_2,$$

where $D_i = \text{not } C_i$, i = 1,2. We've placed underbraces to more easily keep $G$ and $H$ in mind.

Then the events $G$ and $H$ are clearly disjoint; if in a given line of our notebook there is a Yes for $G$, then definitely there will be a No for $H$, and vice versa. So, the *or* in (1.26) become a + in (1.27).

(1.28): The two nodes act physically independently of each other. Thus the events $C_1$ and $C_2$ are stochastically independent, so we applied (1.6). Then we did the same for $D_1$ and $D_2$.

Now, what about $P(X_2 = 2)$? Again, we break big events down into small events, in this case according to the value of $X_1$. Specifically, $X_2 = 2$ if $X_1 = 0$ and $X_2 = 2$ or $X_1 = 1$ and $X_2 = 2$ or $X_1 = 1$ and $X_2 = 2$. Thus, using (1.4), we have

$$
\begin{aligned}
P(X_2 = 2) \quad = \quad & P(X_1 = 0 \text{ and } X_2 = 2) \qquad\qquad (1.30)\\
+ \quad & P(X_1 = 1 \text{ and } X_2 = 2)\\
+ \quad & P(X_1 = 2 \text{ and } X_2 = 2)
\end{aligned}
$$

Since $X_1$ cannot be 0, that first term, $P(X_1 = 0 \text{ and } X_2 = 2)$ is 0. To deal with the second term, $P(X_1 = 1 \text{ and } X_2 = 2)$, we'll use (1.8). Due to the time-sequential nature of our experiment here, it is natural (but certainly not "mandated," as we'll often see other situations) to take $A$ and $B$ in that mailing tube to be $\{X_1 = 1\}$ and $\{X_2 = 2\}$, respectively. So, we write

$$
P(X_1 = 1 \text{ and } X_2 = 2) = P(X_1 = 1)P(X_2 = 2 | X_1 = 1) \qquad (1.31)
$$

To calculate $P(X_1 = 1)$, we use the same kind of reasoning as in Equation (1.25). For the event in question to occur, either node A would send and B wouldn't, or A would refrain from sending and B would send. Thus

$$
P(X_1 = 1) = 2p(1 - p) = 0.48 \qquad (1.32)
$$

Now we need to find $P(X_2 = 2 | X_1 = 1)$. This again involves breaking big events down into small ones. If $X_1 = 1$, then $X_2 = 2$ can occur only if *both* of the following occur:

- Event I: Whichever node was the one to successfully transmit during epoch 1 — and we are given that there indeed was one, since $X_1 = 1$ — now generates a new message.

- Event II: During epoch 2, no successful transmission occurs, i.e., either they both try to send or neither tries to send.

Recalling the definitions of $p$ and $q$ in Section 1.6, we have that

$$
P(X_2 = 2 | X_1 = 1) = q[p^2 + (1 - p)^2] = 0.41 \qquad (1.33)
$$

Thus $P(X_1 = 1 \text{ and } X_2 = 2) = 0.48 \times 0.41 = 0.20$.

We go through a similar analysis for $P(X_1 = 2 \text{ and } X_2 = 2)$: We recall that $P(X_1 = 2) = 0.52$ from before, and find that $P(X_2 = 2 | X_1 = 2) = 0.52$ as well. So we find $P(X_1 = 2 \text{ and } X_2 = 2)$ to be $0.52^2 = 0.27$.

Putting all this together, we find that $P(X_2 = 2) = 0.47$.  This example required a fair amoutn of patience, but the solution patterns used involved the same kind of reasoning as in the bus ridership model earlier.

## 1.7  ALOHA in the Notebook Context

Think of doing the ALOHA "experiment" many, many times.  Let's interpret the numbers we found above, e.g., $P(X_1 = 2) = 0.52$, in the notebook context.

- Run the network for two epochs, starting with both nodes active, the first time, and write the outcome on the first line of the notebook.

- Run the network for two epochs, starting with both nodes active, the second time, and write the outcome on the second line of the notebook.

- Run the network for two epochs, starting with both nodes active, the third time, and write the outcome on the third line of the notebook.

- Run the network for two epochs, starting with both nodes active, the fourth time, and write the outcome on the fourth line of the notebook.

- Imagine you keep doing this, thousands of times, filling thousands of lines in the notebook.

The first seven lines of the notebook might look like Table1.3.  We see that:

- Among those first seven lines in the notebook, 4/7 of them have $X_1 = 2$.  After many, many lines, this fraction will be approximately 0.52.

- Among those first seven lines in the notebook, 3/7 of them have $X_2 = 2$.  After many, many lines, this fraction will be approximately 0.47.[7]

- Among those first seven lines in the notebook, 2/7 of them have $X_1 = 2$ and $X_2 = 2$.  After many, many lines, this fraction will be approximately 0.27.

- Among the first seven lines in the notebook, four of them do not say NA in the $X_2 = 2|X_1 = 2$ column.  **Among these four lines,** two say Yes, a fraction of 2/4.  After many, many lines, this fraction will be approximately 0.52.

---

[7]Don't make anything of the fact that these probabilities nearly add up to 1.

Table 1.3: Top of Notebook for Two-Epoch ALOHA Experiment

| notebook line | $X_1 = 2$ | $X_2 = 2$ | $X_1 = 2$ and $X_2 = 2$ | $X_2 = 2 \mid X_1 = 2$ |
|---|---|---|---|---|
| 1 | Yes | No | No | No |
| 2 | No | No | No | NA |
| 3 | Yes | Yes | Yes | Yes |
| 4 | Yes | No | No | No |
| 5 | Yes | Yes | Yes | Yes |
| 6 | No | No | No | NA |
| 7 | No | Yes | No | NA |

## 1.8   Example: A Simple Board Game

Consider a board game, which for simplicity we'll assume consists of two squares per side, on four sides. The squares are numbered 0-7, and play begins at square 0. A player's token advances according to the roll of a single die. If a player lands on square 3, he/she gets a bonus turn.

Once again: In most problems like this, **it is greatly helpful to first name the quantities or events involved**, and then "translate" English to math. Toward that end, let $R$ denote the player's first roll, and let $B$ be his bonus if there is one, with $B$ being set to 0 if there is no bonus.

Let's find the probability that a player has yet to make a complete circuit of the board—i.e., has not yet reached or passed 0—after the first turn (including the bonus, if any). As usual, we ask "How can the event in question happen?" and we "break big events down into small events." Concerning the latter, we try doing the breakdown according to whether there was a bonus roll:

$$P(\text{doesn't reach or pass } 0) = P(R + B \leq 7)$$

$$
\begin{aligned}
&= \quad P(R \leq 6, R \neq 3 \text{ or } R = 3, B \leq 4) &(1.34)\\
&= \quad P(R \leq 6, R \neq 3) + P(R = 3, B \leq 4) &(1.35)\\
&= \quad P(R \leq 6, R \neq 3) + P(R = 3) \, P(B \leq 4 \mid R = 3)\\
&= \quad \frac{5}{6} + \frac{1}{6} \cdot \frac{4}{6}\\
&= \quad \frac{17}{18} &(1.36)
\end{aligned}
$$

Above we have written commas as a shorthand notation for *and*, a common abbreviation. The reader should supply the reasoning for each of the steps above, citing the relevant mailing tubes.

Note the curious word "try" used above. We said we would *try* breaking down the event of interest according to whether there is a bonus roll. This didn't mean that the subsequent use of mailing tubes might be invalid. They of course *are* valid, but the question is whether breaking down by bonus will help lead us to a solution, as opposed to generating a lot of equations that become more and more difficult to evaluate. In this case, it worked, leading to easily evaluated probabilities, e.g., $P(R = 2) = 1/6$. But if one breakdown approach doesn't work, try another!

Now, here's a shorter way (there are always multiple ways to do a problem):

$$
\begin{aligned}
P(\text{don't reach or pass } 0) &= 1 - P(\text{reach or pass } 0) &(1.37)\\
&= 1 - P(R + B > 7) &(1.38)\\
&= 1 - P(R = 3, B > 4) &(1.39)\\
&= 1 - \frac{1}{6} \cdot \frac{2}{6} &(1.40)\\
&= \frac{17}{18} &(1.41)
\end{aligned}
$$

Now suppose that, according to a telephone report of the game, you hear that on the player's first turn, his token ended up at square 4. Let's find the probability that he got there with the aid of a bonus roll.

**Note that this a conditional probability**—we're finding the probability that the player got a bonus roll, <u>given</u> that we know he ended up at square 4. The word *given* wasn't there in the statement of the problem, but it was implied.

A little thought reveals that we cannot end up at square 4 after making a complete circuit of the board, which simplifies the situation quite a bit.

So, write

$$P(B > 0 \mid R + B = 4)$$

$$= \frac{P(R + B = 4, B > 0)}{P(R + B = 4)}$$

$$= \frac{P(R + B = 4, B > 0)}{P(R + B = 4, B > 0 \text{ or } R + B = 4, B = 0)}$$

$$= \frac{P(R + B = 4, B > 0)}{P(R + B = 4, B > 0) + P(R + B = 4, B = 0)}$$

$$= \frac{P(R = 3, B = 1)}{P(R = 3, B = 1) + P(R = 4)}$$

$$= \frac{\frac{1}{6} \cdot \frac{1}{6}}{\frac{1}{6} \cdot \frac{1}{6} + \frac{1}{6}}$$

$$= \frac{1}{7} \tag{1.42}$$

Again, the reader should make sure to think about which mailing tubes were used in the various steps above, but let's look here at that fourth equality above, as it is a frequent mode of attack in probability problems. In considering the probability $P(R + B = 4, B > 0)$, we ask, what is a simpler—but still equivalent!—description of this event? Well, we see that $R + B = 4, B > 0$ boils down to $R = 3, B = 1$, so we replace the above probability with $P(R = 3, B = 1)$.

Again, this is a very common approach. But be sure to take care that we are in an "if and only if" situation. Yes, $R+B = 4, B > 0$ implies $R = 3, B = 1$, but we must make sure that the converse is true as well. In other words, we must also confirm that $R = 3, B = 1$ implies $R + B = 4, B > 0$. That's trivial in this case, but one can make a subtle error in some problems if one is not careful; otherwise we will have replaced a higher-probability event by a lower-probability one.

# 1.9 Bayes' Rule

## 1.9.1 General Principle

Several of the derivations above follow a common pattern for finding conditional probabilities $P(A|B)$, with the result ending up as

$$P(A|B) = \frac{P(A)P(B|A)}{P(A)P(B|A) + P(\text{not } A)P(B|\text{not } A)} \qquad (1.43)$$

This is known as *Bayes' Theorem* or *Bayes' Rule*. It can be extended easily to cases with several terms in the denominator, arising from situations that need to be broken down into several *disjoint* events $A_1 ..., A_k$ rather than just A and not-A as above:

$$P(A_i|B) = \frac{P(A_i)P(B|A_i)}{\sum_{j=1}^{k} P(A_j)P(B|A_j)} \qquad (1.44)$$

## 1.9.2 Example: Document Classification

Consider an application of the field known as *text classification*. Here we have data on many documents, say articles from the *New York Times*, including their word contents and other characteristics, and wish to have machine determination of the topic categories of new documents.

Say our software sees that a document contains the word *bonds*. Are those financial bonds? Chemical bonds? Parent-child bonds? Maybe the former baseball star Barry Bonds (or even his father, also a pro player, Bobby Bonds)?

Now, what if we also know that the document contains the word *interest*? It's sounding more like a financial document, but on the other hand, the sentence in question could be, say, "Lately there has been much interest among researchers concerning mother-daughter bonds." Then (1.43) could be applied, with $B$ being the event that the document contains the words *bonds* and *interest*, and $A$ being the event that it is a financial ocument.

Those probabilities would have to be estimated from a dataset of documents, known as a *corpus*. For instance, P(financial | bonds, interest) would be estimated as the proportion of financial documents among all documents containing the two specified words.

Note the word *estimated* in the previous paragraph. Even if our corpus is large, it still must be considered only a sample from all *Times* articles. So, there will be some statistical inaccuracy in our estimated probabilities, an issue we will address in the statistics portion of this book; all this starts in Chapter 7.

## 1.10   Random Graph Models

A *graph* consists of *vertices* and *edges*. To understand this, think of a social network. Here the vertices represent people and the edges represent friendships. For the time being, assume that friendship relations are mutual, i.e., if person $i$ says he is friends with person $j$, then $j$ will say the same about $i$.

For any graph, its *adjacency matrix* consists of 1 and 0 entries, with a 1 in row $i$, column $j$ meaning there is an edge from vertex $i$ to vertex $j$. For instance, say we have a simple tiny network of three people, with adjacency matrix

$$\begin{pmatrix} 0 & 1 & 1 \\ 1 & 0 & 0 \\ 1 & 0 & 0 \end{pmatrix} \qquad (1.45)$$

Row 1 of the matrix says that Person 1 is friends with persons 2 and 3, but we see from the other rows that Persons 2 and 3 are not friends with each other.

In any graph, the *degree* of a vertex is its number of edges. So, the degree of vertex $i$ is the number of 1s in row $i$. In the little model above, vertex 1 has degree 2 but the other two vertices each have degree 1.

The assumption that friendships are mutual is described in graph theory as having a *undirected* graph. Note that that implies that the adjacency matrix is symmetric. However, we might model some other networks as *directed*, with adjacency matrices that are not necessarily symmetric. In a large extended family, for example, we could define edges in terms of being an elder sibling; there would be an edge from Person $i$ to Person $j$ if $j$ is an older sibling of $i$.

Graphs need not represent people. They are used in myriad other settings, such as analysis of Web site relations, Internet traffic routing, genetics research and so on.

## 1.10.1   Example: Preferential Attachment Model

A famous graph model is *Preferential Attachment*. Think of it again as an undirected social network, with each edge representing a "friend" relation. The number of vertices grows over time, one vertex per time step. At time 0, we have just two vertices, $v_1$ and $v_2$, with a link between them. At time 1, $v_3$ is added, then $v_4$ at time 2, and so on.

Thus at time 0, each of the two vertices has degree 1. Whenever a new vertex is added to the graph, it randomly chooses an existing vertex to *attach* to, creating a new edge with that existing vertex. The property being modeled is that newcomers tend to attach to the more popular vertices. In making that random choice, the new vertex follows probabilities in proportion to the degrees of the existing edges; the larger the current degree of an existing vertex, the higher the probability that a new vertex will attach to it.

As an example of how the Preferential Attachment Model works, suppose that just before time 2, when $v_4$ is added, the adjacency matrix for the graph is (1.45). Then there will be an edge created between $v_4$ with $v_1$, $v_2$ or $v_3$, with probability 2/4, 1/4 and 1/4, respectively.

Let's find $P(v_4$ attaches to $v_1)$. Let $N_i$ denote the node that $v_i$ attaches to, i = 3,4,... Then, following the solution strategy "break big events down into small events," let's break this question about $v_4$ according to what happens with $v_3$:

$$
\begin{aligned}
P(N_4 = 1) &= P(N_3 = 1 \text{ and } N_4 = 1) + P(N_3 = 2 \text{ and } N_4 = 1) \\
&= P(N_3 = 1)\ P(N_4 = 1 \mid N_3 = 1) + \\
&\quad P(N_3 = 2)\ P(N_4 = 1 \mid N_3 = 2) \\
&= (1/2)(2/4) + (1/2)(1/4) \\
&= 3/8
\end{aligned}
$$

For instance, why is the second term above equal to $(1/2)\ (1/4)$? We are given that $v_3$ had attached to $v_2$, so when $v_4$ comes along, the three existing vertices will have degrees 1, 2 and 1. Thus $v_4$ will attach to them with probabilities 1/4, 2/4 and 1/4, respectively.

# 1.11   Combinatorics-Based Computation

*And though the holes were rather small, they had to count them all* — from
the Beatles song, *A Day in the Life*

In some probability problems all the outcomes are equally likely. The probability computation is then simply a matter of counting all the outcomes
of interest and dividing by the total number of possible outcomes. Of
course, sometimes even such counting can be challenging, but it is simple
in principle. We'll discuss two examples here.

The notation $\binom{n}{k}$ will be used extensively here. It means the number of
ways to choose $k$ things from among $n$, and can be shown to be equal to
$n!/(k!(n-k)!)$.

## 1.11.1   Which Is More Likely in Five Cards, One King or Two Hearts?

Suppose we deal a 5-card hand from a regular 52-card deck. Which is
larger, P(1 king) or P(2 hearts)? Before continuing, take a moment to
guess which one is more likely.

Now, here is how we can compute the probabilities. **The key point is
that all possible hands are equally likely, which implies that all
we need to do is count them.** There are $\binom{52}{5}$ possible hands, so this
is our denominator. For P(1 king), our numerator will be the number of
hands consisting of one king and four non-kings. Since there are four kings
in the deck, the number of ways to choose one king is $\binom{4}{1} = 4$. There are 48
non-kings in the deck, so there are $\binom{48}{4}$ ways to choose them. Every choice
of one king can be combined with every choice of four non-kings, so the
number of hands consisting of one king and four non-kings is the product,
$4 \cdot \binom{48}{4}$. Thus

$$P(1 \text{ king}) = \frac{4 \cdot \binom{48}{4}}{\binom{52}{5}} = 0.299 \qquad (1.46)$$

The same reasoning gives us

$$P(2 \text{ hearts}) = \frac{\binom{13}{2} \cdot \binom{39}{3}}{\binom{52}{5}} = 0.274 \qquad (1.47)$$

So, the 1-king hand is just slightly more likely.

Note again the assumption that all 5-card hands are equally likely. That *is* a realistic assumption, but it's important to understand that it plays a key role here.

By the way, I used the R function **choose()** to evaluate these quantities, running R in interactive mode, e.g.:

```
> choose(13,2) * choose(39,3) / choose(52,5)
[1] 0.2742797
```

R also has a very nice function **combn()** which will generate all the $\binom{n}{k}$ combinations of k things chosen from n, and also will at your option call a user-specified function on each combination. This allows you to save a lot of computational work.

### 1.11.2   Example: Random Groups of Students

A class has 68 students, 48 of whom are computer science majors. The 68 students will be randomly assigned to groups of 4. Find the probability that a random group of 4 has exactly 2 CS majors.

Following the same pattern as above, the probability is

$$\frac{\binom{48}{2}\binom{20}{2}}{\binom{68}{4}}$$

### 1.11.3   Example: Lottery Tickets

Twenty tickets are sold in a lottery, numbered 1 to 20, inclusive. Five tickets are drawn for prizes. Let's find the probability that two of the five winning tickets are even-numbered.

Since there are 10 even-numbered tickets, there are $\binom{10}{2}$ sets of two such tickets. Again as above, we find the desired probability to be

$$\frac{\binom{10}{2}\binom{10}{3}}{\binom{20}{5}} \tag{1.48}$$

Now let's find the probability that two of the five winning tickets are in the range 1 to 5, two are in 6 to 10, and one is in 11 to 20.

Picture yourself picking your tickets. Again there are $\binom{20}{5}$ ways to choose the five tickets. How many of those ways satisfy the stated condition?

Well, first, there are $\binom{5}{2}$ ways to choose two tickets from the range 1 to 5. Once you've done that, there are $\binom{5}{2}$ ways to choose two tickets from the range 6 to 10, and so on. So, The desired probability is then

$$\frac{\binom{5}{2}\binom{5}{2}\binom{10}{1}}{\binom{20}{5}} \tag{1.49}$$

### 1.11.4   Example: Gaps between Numbers

Suppose $m$ numbers are chosen at random, without replacement, from $1, 2, ..., n$. Let $X$ denote the largest gap between consecutive numbers in the chosen set. (Gaps starting at 1 or ending at $n$ don't count unless they are in the chosen set.) For example, if $n = 10$ and $m = 3$, we might choose 2, 6 and 7. The gaps would then be 4 and 1, and $X$ would be 4. Let's write a function to find the exact probability (this is not a simulation) that $X = k$, making use of R's built-in functions **combn()** and **diff()**.

The **diff()** function finds differences between consecutive elements of a vector, e.g.,

```
> diff(c(2,7,18))
[1]   5 11
```

This is exactly what we need:

```
maxgap <- function(n,m,k) {
    tmp <- combn(n,m,checkgap)
    mean(tmp == k)
}


checkgap <- function(cmb) {
    tmp <- diff(cmb)
    max(tmp)
}
```

How does this code work? The call to **combn()** results in the function generating all combinations of the given size and, on each combination, calling **checkgap()**. The results, as seen above, are then stored in the vector **tmp**. Keep in mind, there will be one element in **tmp** for each of the various combinations.

That vector is then used in

```
mean(tmp == k)
```

There are several important issues involved in this seemingly innocuous line of code. In fact, this is a very common pattern in R, so it's important to understand the code fully. Here is what is happening:

The expression **tmp == k** yields a vector (let's call it **u**) of booleans, i.e., TRUEs and FALSEs. The TRUEs arise when a combination yields a max gap of **k**.

So we then find the mean of a vector of booleans. As you may know, in many programming languages, TRUE and FALSE are treated as 1 and 0, respectively. Thus we are applying **mean()** to 1s and 0s.

But an average of 1s and 0s works out to simply the proportion of 1s — which then is the probability value we were seeking! All the combinations are equally likely, so our desired probability is just the proportion of 1s.

## 1.11.5 Multinomial Coefficients

Question: We have a group consisting of 6 Democrats, 5 Republicans and 2 Independents, who will participate in a panel discussion. They will be sitting at a long table. How many seating arrangements are possible, with regard to political affiliation? (So we do not care, for instance, about permuting the individual Democrats within the seats assigned to Democrats.)

Well, there are $\binom{13}{6}$ ways to choose the Democratic seats. Once those are chosen, there are $\binom{7}{5}$ ways to choose the Republican seats. The Independent seats are then already determined, i.e., there will be only way at that point, but let's write it as $\binom{2}{2}$. Thus the total number of seating arrangements is

$$\frac{13!}{6!7!} \cdot \frac{7!}{5!2!} \cdot \frac{2!}{2!0!} \tag{1.50}$$

That reduces to

$$\frac{13!}{6!5!2!} \tag{1.51}$$

The same reasoning yields the following general notion:

*Multinomial Coefficients:* Suppose we have c objects and r bins. Then the number of ways to choose $c_1$ of them to put in bin 1, $c_2$ of them to put in bin 2,..., and $c_r$ of them to put in bin r is

$$\frac{c!}{c_1!...c_r!}, \quad c_1 + ... + c_r = c \tag{1.52}$$

Of course, the "bins" may just be metaphorical. In the political party example above, the "bins " were political parties, and "objects" were seats.

## 1.11.6   Example: Probability of Getting Four Aces in a Bridge Hand

A standard deck of 52 cards is dealt to four players, 13 cards each. One of the players is Millie. What is the probability that Millie is dealt all four aces?

Well, there are

$$\frac{52!}{13!13!13!13!} \tag{1.53}$$

possible deals. (the "objects" are the 52 cards, and the "bins" are the 4 players.) The number of deals in which Millie holds all four aces is the same as the number of deals of 52 - 4 = 48 cards, 13 - 4 = 9 of which go to Millie and 13 each to the other three players, i.e.,

$$\frac{48!}{13!13!13!9!} \tag{1.54}$$

Thus the desired probability is

$$\frac{\frac{48!}{13!13!13!9!}}{\frac{52!}{13!13!13!13!}} = 0.00264 \tag{1.55}$$

## 1.12 Exercises

**Mathematical problems:**

**1.** In the bus ridership example, Section 1.1, say an observer at the second stop notices that no one alights there, but it is dark and the observer couldn't see whether anyone was still on the bus. Find the probability that there was one passenger on the bus at the time.

**2.** In the ALOHA model, Section 1.6, find $P(X_1 = 1 | X_2 = 2)$. Note that there is no issue here with "going backwards in time." The probability here makes perfect sense in the notebook model.

**3.** In the ALOHA model, find $P(X_1 = 2$ or $X_2 = 2)$.

**4.** In general, $P(B \mid A) \neq P(A \mid B)$. Illustrate this using an example with dice, as follows. Let $S$ and $T$ denote the sum and number of even-numbered dice (0, 1 or 2), respectively. Find $P(S = 12 \mid T = 2)$ and $P(T = 2 \mid S = 12)$, and note that they are different.

**5.** Jill arrives daily at a parking lot, at 9:00, 9:15 or 9:30, with probability 0.5, 0.3 and 0.2, respectively. The probability of there being an open space at those times is 0.6, 0.1 and 0.3, respectively.

  (a) Find the probability that she gets a space.

  (b) One day she tells you that she found a space. Determine the most likely time she arrived, and the probability that it was that time.

**6.** Consider the board game example, Section 1.8. Find the probability that after one turn (including bonus, if any), the player is at square 1. Also, find the probability that $B \leq 4$. (Be careful; $B = 0$ does count as $B \leq 4$.)

**7.** Say cars crossing a certain multilane bridge take either 3, 4 or 5 minutes for the trip. 50% take 3 minutes, with a 25% figure each for the 4- and 5-minute trips. We will consider the traversal by three cars, named A, B and C, that simultaneously start crossing the bridge. They are in different lanes, and operate independently.

  (a) Find the probability that the first arrival to the destination is at the 4-minute mark.

  (b) Find the probability that the total trip time for the three cars is 10 minutes.

(c) An observer reports that the three cars arrived at the same time. Find the probability that the cars each took 3 minutes to make the trip.

**8**. Consider the simple ALOHA network model, run for two epochs with $X_0 = 2$. Say we know that there was a total of two transmission attempts. Find the probability that at least one of those attempts occurred during epoch 2. (Note: In the term *attempt*, we aren't distinguishing between successful and failed ones.) Give your analytical answer for general $p$ and $q$.

**9**. Armed with the material on multinomial coefficients in Section 1.11.5, do an alternate calculation of (1.49) .

**10**. Consider the Preferential Attachment Graph model, Section 1.10.1.

(a) Find $P(N_3 = 1 \mid N_4 = 1)$.

(b) Find $P(N_4 = 3)$.

**11**. Consider a three-node version of the ALOHA network example, with all nodes active at time 0. One of the users tells us at the end of epoch 1 that her node was involved in a collision during that epoch. (We have no information from the other two users.) What is the probability that all three nodes were involved in that collision?

**12**. In the random student groups example, Section 1.11.2, suppose there are only 67 students, so that one of the groups will have only 3 students. (Continue to assume there are 48 CS majors.) Say the students are assigned at random to the 17 groups, and then we choose one of the 17 at random. Find the probability that it contains exactly 2 CS students.

**Computational and data problems:**

**13**. Use R's **combn()** function to verify (1.48). The function will walk through each possible subset, and your code can increment a counter for the number of subsets of interest.

**14**. Write an extension of **combn()** that will walk through all possible partitionings, in the context of Section (1.11.5).

**15**. Consider the board game example (but with no bonus rolls). We will be interested in the quantities $t_{ik}$, $0, 1, 2, 4, 5, 6, 7$, the probability that it takes $k$ turns to reach or pass square 0, starting at square i.

Write a recursive function **tik(i,k)** that returns $t_{ik}$. For instance,

```
> tik(1,2)
[1] 0.5833333
> tik(0,2)
[1] 0.4166667
> tik(7,1)
[1] 1
> tik(7,2)
[1] 0
> tik(5,3)
[1] 1
> tik(5,2)
[1] 0.5
> tik(4,4)
[1] 1
> tik(4,3)
[1] 0.5833333
```

**16**. Write a functions with call form

```
permn(x,m,FUN)
```

analogous to **combn()** but for *permutations*. Return value will be a vector or matrix.

Suggestion: Call **combn()** to get each combination, then apply the function **perms()** from e.g., the **partitions** package to generate all permutions corresponding to that combination.

**17**. Apply your **permn()** from Problem 16 to solve the following problem. We choose 8 numbers, $X_1, ..., X_8$ from 1,2,...,50. We are interested in the quantity $W = \Sigma_{i=1}^{7}|X_{i+1} - X_i|$. Find $EW$.

# Chapter 2

# Monte Carlo Simulation

Computer simulation essentially does in actual code what one does conceptually in our "notebook" view of probability (Section 1.2). This is known as *Monte Carlo simulation*.

There are also types of simulation that follow some process in time. One type, *discrete event simulation*, models processes having "discrete" changes, such as a queuing system, in which the number waiting in the queue goes up or down by 1. This is in contrast to, say, modeling the weather, in which temperature and other variables change continuously.

## 2.1   Example: Rolling Dice

If we roll three dice, what is the probability that their total is 8? We could count all the possibilities, or we could get an approximate answer via simulation:

```
# roll d dice; find P(total = k)

probtotk <- function(d,k,nreps) {
    count <- 0
    # do the experiment nreps times -- like doing
    # nreps notebook lines
    for (rep in 1:nreps) {
        sum <- 0
        # roll d dice and find their sum
```

```
      for (j in 1:d) sum <- sum + roll()
      if (sum == k) count <- count + 1
   }
   return(count/nreps)
}

# simulate roll of one die; the possible return
# values are 1,2,3,4,5,6, all equally likely
roll <- function() return(sample(1:6,1))

# example
probtotk(3,8,1000)
```

The call to the built-in R function **sample()** here says to take a sample of
size 1 from the sequence of numbers 1,2,3,4,5,6. That's just what we want
to simulate the rolling of a die. The code

```
for (j in 1:d) sum <- sum + roll()
```

then simulates the tossing of a die *d* times, and computes the sum.

## 2.1.1   First Improvement

Since applications of R often use large amounts of computer time, good R
programmers are always looking for ways to speed things up. Here is an
alternate version of the above program:

```
# roll d dice; find P(total = k)

probtotk <- function(d,k,nreps) {
   count <- 0
   # do the experiment nreps times
   for (rep in 1:nreps)
      total <- sum(sample(1:6,d,replace=TRUE))
      if (total == k) count <- count + 1
   }
   return(count/nreps)
}
```

Let's first discuss the code.

```
sample(1:6,d,replace=TRUE)
```

The call to **sample**() here says, "Generate $d$ random numbers, chosen randomly (i.e., with equal probability) from the integers 1 through 6, with replacement." Well, of course, that simulates tossing the die $d$ times. So, that call returns a $d$-element array, and we then call R's built-in function **sum**() to find the total of the $d$ dice.

This second version of the code here eliminates one explicit loop, which is the key to writing fast code in R. But just as important, it is more compact and clearer in expressing what we are doing in this simulation. The call to R's **sum**() function has both of these properties.

## 2.1.2 Second Improvement

Further improvements are possible. Consider this code:

```
# roll d dice; find P(total = k)

# simulate roll of nd dice; the possible return
# values are 1,2,3,4,5,6, all equally likely
roll <-
    function(nd) return(sample(1:6,nd,replace=TRUE))

probtotk <- function(d,k,nreps) {
    sums <- vector(length=nreps)
    # do the experiment nreps times
    for (rep in 1:nreps) sums[rep] <- sum(roll(d))
    return(mean(sums==k))
}
```

There is quite a bit going on here. This pattern will arise quite often, so let's make sure we have a good command of the details

We are storing the various "notebook lines" in a vector **sums**. We first call **vector**() to allocate space for it.

But the heart of the above code is the expression **sums==k**, which involves the very essence of the R idiom, *vectorization* (Section A.4). At first, the expression looks odd, in that we are comparing a vector **sums**, to a scalar, **k**. But in R, every "scalar" is actually considered a one-element vector.

Fine, **k** is a vector, but wait! It has a different length than **sums**, so how can we compare the two vectors? Well, in R a vector is *recycled*—extended in length, by repeating its values—in order to conform to longer vectors it will be involved with. For instance:

```
> c(2,5) + 4:6
[1]   6 10   8
```

Here we added the vector (2,5) to (4,5,6). The former was first recycled to (2,5,2), resulting in a sum of (6,10,8).[1]

So, in evaluating the expression **sums==k**, R will recycle **k** to a vector consisting of **nreps** copies of **k**, thus conforming to the length of **sums**. The result of the comparison will then be a vector of length **nreps**, consisting of TRUE and FALSE values. In numerical contexts, these are treated at 1s and 0s, respectively. R's **mean()** function will then average those values, resulting in the fraction of 1s! That's exactly what we want.

### 2.1.3   Third Improvement

Even better:

```
roll <- function(nd)
    return(sample(1:6,nd,replace=TRUE))

probtotk <- function(d,k,nreps) {
    # do the experiment nreps times
    sums <- replicate(nreps,sum(roll(d)))
    return(mean(sums==k))
}
```

R's **replicate()** function does what its name implies, in this case executing the call **sum(roll(d))** a total of **nreps** times. That produces a vector, which we then assign to **sums**. And note that we don't have to allocate space for **sums**; **replicate()** produces a vector, allocating space, and then we merely point **sums** to that vector.

The various improvements shown above compactify the code, and in many cases, make it much faster.[2] Note, though, that this comes at the expense of using more memory.

---

[1]There was also a warning message, not shown here. The circumstances under which warnings are or are not generated are beyond our scope here, but recycling is a very common R operation.

[2]You can measure times using R's **system.time()** function, e.g., via the call **system.time(probtotk(3,7,10000))**.

## 2.2   Example: Dice Problem

Suppose three fair dice are rolled. We wish to find the approximate probability that the first die shows fewer than 3 dots, *given* that the total number
of dots for the 3 dice is more than 8, using simulation.

Again, simulation is writing code that implements our "notebook" view of
probability. In this case, we are working with a conditional probability,
which our notebook view defined as follows. $P(B \mid A)$ is the long-run
proportion of the time B occurs, *among those lines in which A occurs*.
Here is the code:

```
dicesim <- function(nreps) {
   count1 <- 0
   count2 <- 0
   for (i in 1:nreps) {
      d <- sample(1:6,3,replace=T)
      # "among those lines in which A occurs"
      if (sum(d) > 8) {
         count1 <- count1 + 1
         if (d[1] < 3) count2 <- count2 + 1
      }
   }
   return(count2 / count1)
}
```

Note carefully that we did NOT use (1.9). That would defeat the purpose
of simulation, which is the model the actual process.

## 2.3   Use of runif() for Simulating Events

To simulate whether a simple event occurs or not, we typically use R function **runif()**. This function generates random numbers from the interval
(0,1), with all the points inside being equally likely. So for instance the
probability that the function returns a value in (0,0.5) is 0.5. Thus here is
code to simulate tossing a coin:

```
if (runif(1) < 0.5)
   heads <- TRUE else heads <- FALSE
```

The argument 1 means we wish to generate just one random number from
the interval (0,1).

## 2.4    Example: Bus Ridership (cont'd.)

Consider the example in Section 1.1. Let's find the probability that after
visiting the tenth stop, the bus is empty. This is too complicated to solve
analytically, but can easily be simulated:

```
nreps <- 10000
nstops <- 10
count <- 0
for (i in 1:nreps) {
   passengers <- 0
   for (j in 1:nstops) {
      if (passengers > 0)   # any alight?
         for (k in 1:passengers)
            if (runif(1) < 0.2)
               passengers <- passengers - 1
      newpass <- sample(0:2,1,prob=c(0.5,0.4,0.1))
      passengers <- passengers + newpass
   }
   if (passengers == 0) count <- count + 1
}
print(count/nreps)
```

Note the different usage of the **sample()** function in the call

```
sample(0:2,1,prob=c(0.5,0.4,0.1))
```

Here we take a sample of size 1 from the set {0,1,2}, but with probabilities
0.5 and so on. Since the third argument for **sample()** is **replace**, not
**prob**, we need to specify the latter in our call.

## 2.5    Example: Board Game (cont'd.)

Recall the board game in Section 1.8. Below is simulation code to find the
probability in (1.42):

```
boardsim <- function(nreps) {
   count4 <- 0
   countbonusgiven4 <- 0
   for (i in 1:nreps) {
      position <- sample(1:6,1)
      if (position == 3) {
```

```
        bonus <- TRUE
        position <-
            (position + sample(1:6,1)) %% 8
    } else bonus <- FALSE
    if (position == 4) {
        count4 <- count4 + 1
        if (bonus) countbonusgiven4 <-
            countbonusgiven4 + 1
    }
}
    return(countbonusgiven4/count4)
}
```

Note the use of R's modulo operator, %%. We are computing board position mod 8, because the position numberts wrap around to 0 after 7.

## 2.6 Example: Broken Rod

Say a glass rod drops and breaks into 5 random pieces. Let's find the probability that the smallest piece has length below 0.02.

First, what does "random" mean here? Let's assume that the 4 break points, treating the left end as 0 and the right end as 1, can be modeled with **runif(4)**. Here then is code to do the job:

```
minpiece <- function(k) {
    breakpts <- sort(runif(k-1))
    lengths <- diff(c(0,breakpts,1))
    min(lengths)
}

# returns the approximate probability
# that the smallest of k pieces will
# have length less than q
bkrod <- function(nreps,k,q) {
    minpieces <- replicate(nreps,minpiece(k))
    mean(minpieces < q)
}

> bkrod(10000,5,0.02)
[1] 0.35
```

So, we generate the break points according to the model, then sort them in order to call R's **diff()** function. (Section 1.11.4.) (Once again, judicious use of R's built-in functions has simplified our code, and speeded it up.) We then find the minimum length.

## 2.7   How Long Should We Run the Simulation?

Clearly, the larger the value of **nreps** in our examples above, the more accurate our simulation results are likely to be. But how large should this value be? Or, more to the point, what measure is there for the degree of accuracy one can expect (whatever that means) for a given value of **nreps**? These questions will be addressed in Chapter 10.

## 2.8   Computational Complements

### 2.8.1   More on the replicate() Function

The call form of **replicate()** is

```
replicate(numberOfReplications,codeBlock)
```

In our example in Section 2.1.3,

```
sums <- replicate(nreps,sum(roll(d)))
```

**codeBlock** was just a single statement, a call to R's **sum()** function. If more than one statement is to be executed, it must be done so in a *block*, a set of statements enclosed by braces, such as, say,

```
f <- function()
{
    replicate(3,
        {
            x <- sample(1:10,5,replace=TRUE)
            range(x)
        }
    )
}
```

## 2.9    Exercises

**Computational and data problems:**

**1.** Modify the simulation code in the broken-rod example, Section 2.6, so that the number of pieces will be random, taking on the values 2, 3 and 4 with probabilities 0.3, 0.3 and 0.4.

**2.** Write code to solve Problem 11, Chapter 1.

**3.** Write a function with call form **paSim(ngen)** that simulates **ngen** generations of the Preferential Attachment Model, Section 1.10.1. It will return the adjacency matrix, with **ngen** rows and columns. Use this code to find the approximate probability that $v_1$ has degree 3 after $v_5$ joins the network.

**4.** Modify the simulation of the board game example in Section 2.5 to incorporate a random starting point, which we take to be squares 0 to 7 with probability 1/8 each. Also, add code to find $P(X = 7)$, where $X$ is the position after one turn (including bonus, if any).

**5.** Say we toss a coin until we obtain $k$ consecutive heads. Write a function with call form **ngtm(k,m,nreps)** that uses simulation to find and return the approximate probability that it takes more than **m** tosses to achieve the goal.

**6.** Alter the code in Section 2.4 to find the probability that the bus will be empty when arriving to *at least* one stop among the first 10.

**7.** Consider the typical loop line in the simulation code we've seen here:

```
for (rep in 1:nreps) {
```

The larger the value of **nreps**, the more likely our result is of high accuracy. We will discuss this point more precisely later, but for now, one way to assess whether our **nreps** value is large enough would be to see whether things have stabilized, as follows.

Alter the code in Section 2.1 so that it plots the values of **count / i** for every tenth value of **i**. Has the curve mostly leveled off? (You'll need to read ahead Section 5.7.1.)

**8.** The code in Section 1.11.4 finds exact probabilities. But for larger $n$ and $k$, the enumeration of all possible combinations would be quite time-consuming. Convert the function **maxgap()** to simulation.

# Chapter 3

# Discrete Random Variables: Expected Value

This and the next chapter will introduce entities called *discrete random variables*. Some properties will be derived for means of such variables, with most of these properties actually holding for random variables in general. Well, all of that seems abstract to you at this point, so let's get started.

## 3.1   Random Variables

In a more mathematical formulation, with a formal sample space defined, a random variable would be defined to be a real-valued function whose domain is the sample space. But again, we take a more intuitive approach here.

**Definition 3** *A random variable is a numerical outcome of our experiment.*

For instance, consider our old example in which we roll two dice, with $X$ and $Y$ denoting the number of dots we get on the blue and yellow dice, respectively. Then $X$ and $Y$ are random variables, as they are numerical outcomes of the experiment. Moreover, $X + Y$, $2XY$, $\sin(XY)$ and so on are also random variables.

## 3.2   Discrete Random Variables

In our dice example, the random variable $X$ could take on six values in the set {1,2,3,4,5,6}. We say that the *support* of $X$ is {1,2,3,4,5,6}, meaning the list of the values the random variable can take on. This is a finite set here.

In the ALOHA example, Section 1.6, $X_1$ and $X_2$ each have support {0,1,2}, again a finite set.

Now think of another experiment, in which we toss a coin until we get a head. Let $N$ be the number of tosses needed. Then the support of $N$ is the set {1,2,3,...} This is a countably infinite set.[1]

Now think of one more experiment, in which we throw a dart at the interval (0,1), and assume that the place that is hit, R, can take on any of the values between 0 and 1. Here the support is an uncountably infinite set.

We say that $X$, $X_1$, $X_2$ and $N$ are *discrete* random variables, while R is *continuous*. We'll discuss continuous random variables in Chapter 6.

Note that discrete random variables are not necessarily integer-valued. Consider the random variable $X$ above (number of dots showing on a die). Define $W = 0.1X$. $W$ still takes on values in a finite set (0, 0.1,...,0.6), so it too is discrete.

## 3.3   Independent Random Variables

We already have a definition for the independence of events; what about independence of random variables? The answer is that we say two *random variables* are independent if *events* corresponding to them are independent.

In the dice example above, it is intuitively clear that the random variables $X$ and $Y$ "do not affect" each other. If I know, say, that $X = 6$, that knowledge won't help me guess $Y$ at all. For instance, the probability that $Y = 2$, knowing $X$, is still 1/6. Writing this mathematically, we have

$$P(Y = 2 \mid X = 6) = P(Y = 2) \tag{3.1}$$

---

[1]This is a concept from the fundamental theory of mathematics. Roughly speaking, it means that the set can be assigned an integer labeling, i.e., item number 1, item number 2 and so on. The set of positive even numbers is countable, as we can say 2 is item number 1, 4 is item number 2 and so on. It can be shown that even the set of all rational numbers is countable.

which in turn implies

$$P(Y = 2 \text{ and } X = 6) = P(Y = 2) \, P(X = 6) \qquad (3.2)$$

In other words, the events $\{X = 6\}$ and $\{Y = 2\}$ are independent, and similarly the events $\{X = i\}$ and $\{Y = j\}$ are independent for any $i$ and $j$. This leads to our formal definition of independence:

**Definition 4** *Random variables $X$ and $Y$ are said to be independent if for any sets $I$ and $J$, the corresponding events $\{X$ is in $I\}$ and $\{Y$ is in $J\}$ are independent, i.e.,*

$$P(X \text{ is in } I \text{ and } Y \text{ is in } J) = P(X \text{ is in } I) \cdot P(Y \text{ is in } J) \qquad (3.3)$$

So the concept simply means that $X$ doesn't affect $Y$ and vice versa, in the sense that knowledge of one does not affect probabilities involving the other. The definition extends in the obvious way to sets of more than two random variables.

The notion of independent random variables is absolutely central to the field of probability and statistics, and will pervade this entire book.

## 3.4 Example: The Monty Hall Problem

This problem, while quite simply stated, has a reputation as being extremely confusing and difficult to solve [37]. Yet it is actually an example of how the use of random variables in "translating" the English statement of a probability problem to mathematical terms can simplify and clarify one's thinking, making the problem easier to solve. This "translation" process consists simply of naming the quantities. You'll see that here with the Monty Hall Problem.

**Imagine, this simple device of introducing named random variables into our analysis makes a problem that has vexed famous mathematicians quite easy to solve!**

The Monty Hall Problem, which gets its name from a popular TV game show host, involves a contestant choosing one of three doors. Behind one door is a new automobile, while the other two doors lead to goats. The contestant chooses a door and receives the prize behind the door.

The host knows which door leads to the car. To make things interesting, after the contestant chooses, the host will open one of the other doors not chosen, showing that it leads to a goat. Should the contestant now change her choice to the remaining door, i.e., the one that she didn't choose and the host didn't open?

Many people answer No, reasoning that the two doors not opened yet each have probability 1/2 of leading to the car. But the correct answer is actually that the remaining door (not chosen by the contestant and not opened by the host) has probability 2/3, and thus the contestant should switch to it. Let's see why.

Again, **the key is to name some random variables.** Let

- $C$ = contestant's choice of door (1, 2 or 3)

- $H$ = host's choice of door (1, 2 or 3), after contestant chooses

- $A$ = door that leads to the automobile

We can make things more concrete by considering the case $C = 1$, $H = 2$. The mathematical formulation of the problem is then to find the probability that the contestant should change her mind, i.e., the probability that the car is actually behind door 3:

$$P(A = 3 \mid C = 1, \ H = 2) = \frac{P(A = 3, \ C = 1, \ H = 2)}{P(C = 1, \ H = 2)} \qquad (3.4)$$

**You may be amazed to learn that, really, we are already done with the hard part of the problem.** Writing down (3.4) was the core of the solution, and all that remains is to calculate the various quantities above. This will take a while, but it is pretty mechanical from here on, simply going through steps like those we took so often in earlier chapters.

Write the numerator as

$$P(A = 3, \ C = 1) \ P(H = 2 \mid A = 3, \ C = 1) \qquad (3.5)$$

Since $C$ and $A$ are independent random variables, the value of the first factor in (3.5) is

$$\frac{1}{3} \cdot \frac{1}{3} = \frac{1}{9} \qquad (3.6)$$

What about the second factor? Remember, in calculating $P(H = 2 \mid A = 3, C = 1)$, we are given in that case that the host knows that $A = 3$, and since the contestant has chosen door 1, the host will open the only remaining door that conceals a goat, i.e., door 2. In other words,

$$P(H = 2 \mid A = 3, C = 1) = 1 \tag{3.7}$$

Now consider the denominator in (3.4). We can, as usual, "break big events down into small events." For the breakdown variable, it seems natural to use $A$, so let's try that one:

$$P(C = 1, H = 2) = P(A = 3, C = 1, H = 2) + P(A = 1, C = 1, H = 2) \tag{3.8}$$

(There is no $A = 2$ case, as the host, knowing the car is behind door 2, wouldn't choose it.)

We already calculated the first term. Let's look at the second, which is equal to

$$P(A = 1, C = 1) \, P(H = 2 \mid A = 1, C = 1) \tag{3.9}$$

If the host knows the car is behind door 1 and the contestant chooses that door, the host would randomly choose between doors 2 and 3, so

$$P(H = 2 \mid A = 1, C = 1) = \frac{1}{2} \tag{3.10}$$

Meanwhile, similar to before,

$$P(A = 1, C = 1) = \frac{1}{3} \cdot \frac{1}{3} = \frac{1}{9} \tag{3.11}$$

So, altogether we have

$$P(A = 3 \mid C = 1, H = 2) = \frac{\frac{1}{9} \cdot 1}{\frac{1}{9} \cdot 1 + \frac{1}{9} \cdot \frac{1}{2}} = \frac{2}{3} \tag{3.12}$$

Even Paul Erdös, one of the most famous mathematicians in history, is said to have given the wrong answer to this problem. Presumably he would have

avoided this by writing out his analysis in terms of random variables, as
above, rather than say, a wordy, imprecise and ultimately wrong solution.

## 3.5   Expected Value

### 3.5.1   Generality — Not Just for Discrete Random Variables

The concepts and properties introduced in this section form the very core
of probability and statistics. **Except for some specific calculations,
these apply to both discrete and continuous random variables,
and even the exceptions will be analogous.**

The properties developed for *variance*, defined later, also hold for both
discrete and continuous random variables.

### 3.5.2   Misnomer

The term "expected value" is one of the many misnomers one encounters
in tech circles. The expected value is actually not something we "expect"
to occur. On the contrary, it's often pretty unlikely or even impossible.

For instance, let $H$ denote the number of heads we get in tossing a coin 1000
times. The expected value, you'll see later, is 500. This is not surprising,
given the symmetry of the situation and the fact (to be brought in shortly)
that the expected value is the mean. But $P(H = 500)$ turns out to be
about 0.025. In other words, we certainly should not "expect" $H$ to be
500.

Of course, even worse is the example of the number of dots that come up
when we roll a fair die. The expected value will turn out to be 3.5, a value
which not only rarely comes up, but in fact never does.

In spite of being misnamed, expected value plays an absolutely central role
in probability and statistics.

### 3.5.3   Definition and Notebook View

**Definition 5** *Consider a repeatable experiment with random variable $X$.
We say that the expected value of $X$ is the long-run average value of $X$, as*

*we repeat the experiment indefinitely.*

In our notebook, there will be a column for $X$. Let $X_i$ denote the value of $X$ in the $i^{th}$ row of the notebook. Then the long-run average of $X$, i.e., the long-run average in the $X$ column of the notebook, is[2]

$$\lim_{n\to\infty} \frac{X_1 + ... + X_n}{n} \qquad (3.13)$$

To make this more explicit, look at the partial notebook example in Table 3.1. Here we roll two dice, and let $S$ denote their sum. $E(S)$ is then the long-run average of the values in the $S$ column.

Due to the long-run average nature of $E()$, we often simply call it the mean.

## 3.6 Properties of Expected Value

Here we will derive a handy computational formula for the expected value of a discrete random variable, and derive properties of the concept. **You will be using these throughout the remainder of the book, so take extra time here.**

### 3.6.1 Computational Formula

Equation (3.13) defined expected value, but one almost never computes it directly from the definition. Instead, we use a formula, which we will now derive.

Suppose for instance our experiment is to toss 10 coins. Let $X$ denote the number of heads we get out of 10. We might get four heads in the first repetition of the experiment, i.e., $X_1 = 4$, seven heads in the second repetition, so $X_2 = 7$, and so on. Intuitively, the long-run average value of $X$ will be 5. (This will be proven below.) Thus we say that the expected value of $X$ is 5, and write $E(X) = 5$. But let's confirm that, and derive a key formula in the process.

---

[2]The above definition puts the cart before the horse, as it presumes that the limit exists. Theoretically speaking, this might not be the case. However, it does exist if the $X_i$ have finite lower and upper bounds, which is always true in the real world. For instance, no person has height of 50 feet, say, and no one has negative height either. In this book, we will usually speak of "the" expected value of a random variable without adding the qualifier "if it exists."

Now let $K_{in}$ be the number of times the value $i$ occurs among $X_1, ..., X_n$, $i = 0, ..., 10$, $n = 1, 2, 3, ...$ For instance, $K_{4,20}$ is the number of times we get four heads, in the first 20 repetitions of our experiment. Then

$$
\begin{aligned}
E(X) \quad &= \quad \lim_{n \to \infty} \frac{X_1 + ... + X_n}{n} && (3.14) \\
&= \quad \lim_{n \to \infty} \frac{0 \cdot K_{0n} + 1 \cdot K_{1n} + 2 \cdot K_{2n} + ... + 10 \cdot K_{10,n}}{n} && (3.15) \\
&= \quad \sum_{i=0}^{10} i \cdot \lim_{n \to \infty} \frac{K_{in}}{n} && (3.16)
\end{aligned}
$$

To understand that second equation, suppose when $n = 5$, i.e., after the fifth line of our notebook, we have 2, 3, 1, 2 and 1 for our values of $X_1, X_2, X_3, X_4, X_5$. Then we can group the 2s together and group the 1s together, and write

$$
2 + 3 + 1 + 2 + 1 = 2 \times 2 + 2 \times 1 + 1 \times 3 \qquad (3.17)
$$

We have two 2s, so $K_{2,5} = 2$ and so on.

But $\lim_{n \to \infty} K_{in}/n$ is the long-run fraction of the time that $X = i$. In other words, it's $P(X = i)$! So,

$$
E(X) = \sum_{i=0}^{10} i \cdot P(X = i) \qquad (3.18)
$$

So in general we have:

**Property A:**

The expected value of a discrete random variable $X$ which has support $A$ is

$$
E(X) = \sum_{c \in A} c \, P(X = c) \qquad (3.19)
$$

We'll use the above formula quite frequently, but it is worth rewriting it a bit:

$$
E(X) = \sum_{c \in A} P(X = c) \, c \qquad (3.20)
$$

The probabilities $P(X = c)$ are of course numbers in $[0,1]$. So, we see that **E(X) amounts to a weighted sum of the values in the support of X, with the weights being the probabilities of those values.**

As mentioned, (3.19) is the formula we'll usually use in computing expected value. The preceding equations were derivation, to motivate the formula.

Note again that (3.19) is not the *definition* of expected value; that was in (3.13). It is quite important to distinguish between the two.[3] The definition is important for our intuitive understanding, while the formula is what we will turn to when actually computing expected values.

By the way, note the word *discrete* above. We will see later that for the case of continuous random variables, the sum in (3.19) will become an integral.

Now, here are a couple of examples of the formula in action. First, the coin toss example from above. It will be shown in Section 5.4.2 that in our example above in which $X$ is the number of heads we get in 10 tosses of a coin,

$$P(X = i) = \binom{10}{i} 0.5^i (1 - 0.5)^{10-i} \qquad (3.21)$$

So

$$E(X) = \sum_{i=0}^{10} i \binom{10}{i} 0.5^i (1 - 0.5)^{10-i} \qquad (3.22)$$

(It is customary to use capital letters for random variables, e.g., $X$ here, and lower-case letters for values taken on by a random variable, e.g., $i$ here.)

After evaluating the sum, we find that $E(X) = 5$, as promised.

For $X$, the number of dots we get in one roll of a die,

$$E(X) = \sum_{c=1}^{6} c \cdot \frac{1}{6} = 3.5 \qquad (3.23)$$

By the way, it is also customary to write $EX$ instead of $E(X)$, whenever removal of the parentheses does not cause any ambiguity. An example in

---

[3]The matter is made a little more confusing by the fact that many books do in fact treat (3.19) as the definition, with (3.13) being the consequence.

which it would produce ambiguity is $E(U^2)$. The expression $EU^2$ might be taken to mean either $E(U^2)$, which is what we want, or $(EU)^2$, which is not what we want. But if we simply want $E(U)$, then writing $EU$ causes no problem.

Again consider our dice example, with $X$ and $Y$ denoting the number of dots on the yellow and blue die, respectively. Write the sum as $S = X + Y$. First note that the support of $S$ is $\{2, 3, 4, ..., 12\}$. Thus in order to find $ES$, we need to find $P(S = i)$, $i = 2, 3, ..., 12$. But these probabilities are straightforward. For instance, for $P(S = 3)$, just note that $S$ can be 3 if either $X = 1$ and $Y = 2$ or vice versa, for a total probability of 2/36. So,

$$E(S) = 2 \cdot \frac{1}{36} + 3 \cdot \frac{2}{36} + 4 \cdot \frac{3}{36} + ...12 \cdot \frac{1}{36} = 7 \qquad (3.24)$$

In example in which $N$ was the number of coin tosses to obtain a head,

$$E(N) = \sum_{c=1}^{\infty} c \cdot \frac{1}{2^c} = 2 \qquad (3.25)$$

(We will not go into the details here concerning how the sum of this particular infinite series can be evaluated. See Section 5.4.1.)

## 3.6.2   Further Properties of Expected Value

We found above in (3.24) that with $S = X + Y$, we have $E(S) = 7$. This means that in the long-run average in column $S$ in Table 3.1 is 7.

But we really could have deduced that without the computation in (3.24), as follows. Since the $S$ column is the sum of the $X$ and $Y$ columns, the long-run average in the $S$ column must be the sum of the long-run averages of the $X$ and $Y$ columns. Since those two averages are each 3.5, we must have $ES = 7$. In other words:

**Property B:**

For any random variables $U$ and $V$, the expected value of a new random variable $D = U + V$ is the sum of the expected values of $U$ and $V$:

$$E(U + V) = E(U) + E(V) \qquad (3.26)$$

Note carefully that $U$ and $V$ do NOT need to be independent random

Table 3.1: Expanded Notebook for the Dice Problem

| notebook line | $X$ | $Y$ | $S$ |
|---|---|---|---|
| 1 | 2 | 6 | 8 |
| 2 | 3 | 1 | 4 |
| 3 | 1 | 1 | 2 |
| 4 | 4 | 2 | 6 |
| 5 | 1 | 1 | 2 |
| 6 | 3 | 4 | 7 |
| 7 | 5 | 1 | 6 |
| 8 | 3 | 6 | 9 |
| 9 | 2 | 5 | 7 |

variables for this relation to hold. You should convince yourself of this fact intuitively **by thinking about the notebook notion**, as in the $S$ example above.

While you are at it, use the notebook notion to convince yourself of the following:

**Properties C:**

- For any random variable $U$ and constant $a$, then

$$E(aU) = a \ EU \qquad (3.27)$$

Again, $aU$ is a new random variable, defined in terms of an old one. (3.27) shows how to get the new expected value.

- For random variables $X$ and $Y$ — not necessarily independent — and constants $a$ and $b$, we have

$$E(aX + bY) = a \ EX + b \ EY \qquad (3.28)$$

This follows by taking $U = aX$ and $V = bY$ in (3.26), and then using (3.27).

By induction, for constants $a_1, ..., a_k$ and random variables $X_1, ..., X_k$, form the new random variable $a_1 X_1 + ... + a_k X_k$. Then

$$E(a_1 X_1 + ... + a_k X_k) = a_1 E X_1 + ... + a_k E X_k \qquad (3.29)$$

- For any constant $b$, we have

$$E(b) = b \qquad (3.30)$$

This should make sense. If the "random" variable $X$ has the constant value 3, say, then the $X$ column in the notebook will consist entirely of 3s. Thus the long-run average value in that column will be 3, so $EX = 3$.

For instance, say U is temperature in Celsius. Then the temperature in Fahrenheit is $W = \frac{9}{5}U + 32$. So, W is a new random variable, and we can get its expected value from that of $U$ by using (3.28); we take $X$ and $Y$ to be $U$ and 1, with $a = \frac{9}{5}$ and b = 32.

Now, to introduce the next property, consider a function $g()$ of one variable, and let $W = g(X)$. $W$ is then a random variable too. Say $X$ has support $A$, as in (3.19). Then $W$ has support $B = \{g(c) : c \in A\}$. (There may be some repeated values in $A$, as seen in the small example below.)

For instance, say $g()$ is the squaring function, and $X$ takes on the values -1, 0 and 1, with probability 0.5, 0.4 and 0.1. Then

$$A = \{-1, 0, 1\} \qquad (3.31)$$

and

$$B = \{0, 1\} \qquad (3.32)$$

Now, by (3.19),

$$EW = \sum_{d \in B} d \cdot P(W = d) \qquad (3.33)$$

But we can translate (3.33) to terms of $X$:

**Property D:**

$$E[g(X)] = \sum_{c \in A} g(c) \cdot P(X = c) \tag{3.34}$$

where the sum ranges over all values $c$ that can be taken on by $X$.

For example, suppose for some odd reason we are interested in finding $E(\sqrt{X})$, where $X$ is the number of dots we get when we roll one die. Let $W = \sqrt{X}$. Then $W$ is another random variable, and is discrete, since it takes on only a finite number of values. (The fact that most of the values are not integers is irrelevant.) We want to find $EW$.

Well, $W$ is a function of $X$, with $g(t) = \sqrt{t}$. So, (3.34) tells us to make a list of values in the support of $X$, i.e., 1,2,3,4,5,6, and a list of the corresponding probabilities for $X$, which are all $\frac{1}{6}$. Substituting into (3.34), we find that

$$E(\sqrt{X}) = \sum_{i=1}^{6} \sqrt{i} \cdot \frac{1}{6} \approx 1.81 \tag{3.35}$$

(The above sum can be evaluated compactly in R as sum(sqrt(1:6))/6.)

**Note:** Equation (3.34) will be one of the most heavily used formulas in this book. Make sure you keep it in mind.

**Property E:** If $U$ and $V$ are independent, then

$$E(UV) = EU \cdot EV \tag{3.36}$$

In the dice example, for instance, let $D$ denote the product of the numbers of blue dots and yellow dots, i.e., $D = XY$. Then since we found earlier that $EX = EY = 3.5$,

$$E(D) = 3.5^2 = 12.25 \tag{3.37}$$

Note that we do need $U$ and $V$ to be independent here, in contrast to Property B. Unfortunately, Equation (3.36) doesn't have an easy "notebook proof." A formal one is given in Section 3.11.1.

**The properties of expected discussed above are key to the entire remainder of this book. You should notice immediately when you are in a setting in which they are applicable. For instance, if you**

see the expected value of the sum of two random variables, you
should instinctively think of Property B right away.


## 3.7   Example: Bus Ridership

In Section 1.1, let's find the expected value of $L_1$, the number of passengers
on the bus as it leaves the first stop.

To use (3.19), we need $P(L_1 = i)$ for all $i$ in the support of $L_1$. But since
the bus arrives empty to the first stop, $L_1 = B_1$ (recall that the latter is
the number who board at the first stop). The support of $B_1$ is 0, 1 and 2,
which it takes on with probabilities 0.5, 0.4 and 0.1. So,

$$EL_1 = 0.5(0) + 0.4(1) + 0.1(2) = 0.6 \tag{3.38}$$

If we observe the bus on many, many days, on average it will leave the first
stop with 0.6 passengers.

What about $EL_2$? Here the support is $\{0,1,2,3,4\}$, so we need $P(L_2 = i)$, $i = 0,1,2,3,4$. We already found in Section 1.5 that $P(L_2 = 0) = 0.292$.
The other probabilities are found in a similar manner.


## 3.8   Example: Predicting Product Demand

Prediction is a core area of data science. We will study it in detail in
Chapter 15, but let's consider a very simple model now.

Let $D_i$, $i = 1,2,...$ denote the number of items of a certain kind sold on
days 1,2,..., with the support of each day's sales being $\{1,2,3\}$. Suppose
data show that if a day's demand is 1 or 2, the next day's sales will be
1, 2 or 3 with probability 1/3 each. But on high-demand days, i.e., those
on which 3 items are sold, the number sold the next day will be 1, 2 or
3, with probability 0.2, 0.2 and 0.6, respectively; in other words, the high
demand has "momentum," with one 3-item day more likely to be followed
by another.

Say today 3 items were sold. Findiing the expected number for tomorrow
is straightforward:

$$0.2(1) + 0.2(2) + 0.6(3) = 2.4 \tag{3.39}$$

But what should our forecast be for $M$, the sales *two* days from now? Again, the support is $\{1,2,3\}$, but the probabilities $P(M = i)$ are different. For instance, what about $P(M = 3)$?

Once again, "break big events into small events," in this case breaking down by whether tomorrow is another high-demand day, resulting in a sum of two terms:

$$P(M = 3) = 0.6 \times 0.6 + 0.4 \times 1/3 \qquad (3.40)$$

or about 0.4933.

## 3.9 Expected Values via Simulation

For expected values $EX$ that are too complex to find analytically, simulation provides an alternative. Follwing the definition of expected value as the long-run average, we simply simulate **nreps** replications of the experiment, record the value of $X$ in each one, and output the average of those **nreps** values.

Here is a modified version of the code in Section 2.4, to find the approximate value of the expected number of passengers on the bus as it leaves the tenth stop:

```
nreps <- 10000
nstops <- 10
total <- 0
for (i in 1:nreps) {
   passengers <- 0
   for (j in 1:nstops) {
      if (passengers > 0)
         for (k in 1:passengers)
            if (runif(1) < 0.2)
               passengers <- passengers - 1
      newpass <- sample(0:2,1,prob=c(0.5,0.4,0.1))
      passengers <- passengers + newpass
   }
   total <- total + passengers
}
print(total/nreps)
```

We keep a running total of the number passengers at stop 10 for each repetition, then divide by **nreps** to obtain the long-run average.

## 3.10   Casinos, Insurance Companies and "Sum Users," Compared to Others

The expected value is intended as a *measure of central tendency*, also called a *measure of location*, i.e., as some sort of definition of the probablistic "middle" in the range of a random variable. There are various other such measures one can use, such as the *median*, the halfway point of a distribution (0.5 probability below, 0.5 above), and today they are recognized as being superior to the mean in certain senses. For historical reasons, the mean continues to play an absolutely central role in probability and statistics, yet one should understand its limitations. (This discussion will be general, not limited to discrete random variables.)

(**Warning:** The concept of the mean is likely so ingrained in your consciousness that you simply take it for granted that you know what the mean means, no pun intended. But try to take a step back, and think of the mean afresh in what follows.)

It's clear that the mean is terribly overused. Consider, for example, an attempt to describe how wealthy (or not) people are in the city of Davis. If suddenly billionaire Bill Gates were to move into town, that would skew the value of the mean beyond recognition.

But even without Mr. Gates, there is a question as to whether the mean has that much meaning. After all, what is so meaningful about summing our data and dividing by the number of data points? By contrast, the median has an easy intuitive meaning. But although the mean has familiarity, one would be hard pressed to justify it as a measure of central tendency.

What, for example, does Equation (3.13) mean in the context of people's heights in Davis? We would sample a person at random and record his/her height as $X_1$. Then we'd sample another person, to get $X_2$, and so on. Fine, but in that context, what would (3.13) mean? The answer is, not much. So the significance of the mean height of people in Davis would be hard to explain.

For a casino, though, (3.13) means plenty. Say $X$ is the amount a gambler wins on a play of a roulette wheel, and suppose (3.13) is equal to $1.88. Then after, say, 1000 plays of the wheel (not necessarily by the same gambler), the casino knows from (3.13) that it will have paid out a total of about $1,880. So if the casino charges, say $1.95 per play, it will have made a profit of about $70 over those 1000 plays. It might be a bit more or less than that amount, but the casino can be pretty sure that it will be around $70, and they can plan their business accordingly.

The same principle holds for insurance companies, concerning how much they pay out in claims — another quantity that comes in the form of a sum. With a large number of customers, they know ("expect"!) approximately how much they will pay out, and thus can set their premiums accordingly. Here again the mean has a tangible, practical meaning.

The key point in the casino and insurance companies examples is that they are interested in *totals*, such as *total* payouts on a blackjack table over a month's time, or *total* insurance claims paid in a year. Another example might be the number of defectives in a batch of computer chips; the manufacturer is interested in the *total* number of defectives chips produced, say in a month. Since the mean is by definition a *total* (divided by the number of data points), the mean will be of direct interest to casinos etc.

For general applications, such as studying the distribution of heights in Davis, totals are not of inherent interest, and thus the use of the mean is questionable. Nevertheless, the mean has certain mathematical properties, such as (3.26), that have allowed the rich development of the fields of probability and statistics over the years. The median, by contrast, does not have nice mathematical properties. In many cases, the mean won't be too different from the median anyway (barring Bill Gates moving into town), so you might think of the mean as a convenient substitute for the median. The mean has become entrenched in statistics, and we will use it often.

# 3.11   Mathematical Complements

## 3.11.1   Proof of Property E

Let $A_U$ and $A_V$ be the supports of $U$ and $V$. Since $UV$ is a discrete random variable, we can use (3.19). To find $E(UV)$, we multiply each value in its support by the probability of that value:

$$E(UV) \;=\; \sum_{i \text{ in } A_U} \sum_{j \text{ in } A_V} ij \, P(U = i \text{ and } V = j) \qquad (3.41)$$

$$=\; \sum_{i \text{ in } A_U} \sum_{j \text{ in } A_V} ij \, P(U = i) \, P(V = j) \qquad (3.42)$$

$$=\; \sum_{i \text{ in } A_U} i \, P(U = i) \sum_{j \text{ in } A_V} j \, P(V = j) \qquad (3.43)$$

$$=\; EU \cdot EV \qquad (3.44)$$

That first equation is basically (3.19). We then use the definition of independence, aod factor constants in $i$ out of the sum over $j$.

## 3.12   Exercises

**Mathematical problems:**

**1.** In Section 3.7, finish the calculation of $EL_2$.

**2.** In Section 3.8, finish the calculation of $EM$.

**3.** Consider the ALOHA example, Section 1.6, using two nodes, both of which start out active, with $p = 0.4$ and $q = 0.8$. Find the expected value of the number of attempted transmissions (successful or not) during the first epoch.

**4.** In the student groups example, Section 1.11.2, find the expected number of computer science students in the three-person group.

**5.** In Exercise 12, Chapter 1, find the expected number of computer science students in the three-person group.

**6.** Four players are dealt bridge hands (Section 1.11.6). Some may have no aces. Find the expected number of players having no aces.

**Computational and data problems:**

**7.** Consider the code in Section 2.5. Extend it to find the expected number of turns until reaching or passing 0.

**8.** Say a game consists of rolling a die until the player accumulates 15 dots. Write simulation code to find the expected number of rolls needed to win.

**9.** Again assume the setting in Exercise 3. Use simulation to find the expected time (number of epochs) needed for both original messages to get through.

**10**. Modify the code in Section 1.11.4 to write a function with call form **gapsSim(n,m)** to find the expected size of the largest gap. (Note that this is NOT a simulation, as you are enumerating all possiblities.)

# Chapter 4

# Discrete Random Variables: Variance

Continuing from the last chapter, we extend the notion of expected value to *variance* Again, most of the properties derived here will actually hold for random variables in general, which we will discuss in later chapters.

## 4.1  Variance

As in Section 3.5, the concepts and properties introduced in this section form the very core of probability and statistics. **Except for some specific calculations, these apply to both discrete and continuous random variables.**

### 4.1.1  Definition

While the expected value tells us the average value a random variable takes on, we also need a measure of the random variable's variability — how much does it wander from one line of the notebook to another? In other words, we want a measure of *dispersion*. The classical measure is *variance*, defined to be the mean squared difference between a random variable and its mean:

**Definition 6** *For a random variable U for which the expected values writ-*

Table 4.1: Notebook view of variance

| line | X | W |
|------|---|------|
| 1 | 2 | 2.25 |
| 2 | 5 | 2.25 |
| 3 | 6 | 6.25 |
| 4 | 3 | 0.25 |
| 5 | 5 | 2.25 |
| 6 | 1 | 6.25 |

*ten below exist, the variance of U is defined to be*

$$Var(U) = E[(U - EU)^2] \tag{4.1}$$

*The square root of the variance is called the standard deviation.*

For $X$, the number of dots obtained in one roll of a die, we know from the last chapter that $EX = 3.5$, so the variance of $X$ would be

$$Var(X) = E[(X - 3.5)^2] \tag{4.2}$$

Remember what this means: We have a random variable $X$, and we're creating a new random variable, $W = (X - 3.5)^2$, which is a function of the old one. We are then finding the expected value of that new random variable $W$.

In the notebook view, $E[(X - 3.5)^2]$ is the long-run average of the $W$ column, as seen in Table 4.1. To evaluate this, apply (3.34) with $g(c) = (c - 3.5)^2$:

$$Var(X) = \sum_{c=1}^{6} (c - 3.5)^2 \cdot \frac{1}{6} = 2.92 \tag{4.3}$$

You can see that variance does indeed give us a measure of dispersion. In the expression $Var(U) = E[(U - EU)^2]$, if the values of U are mostly clustered near its mean, then $(U - EU)^2$ will usually be small, and thus the variance of U will be small; if there is wide variation in U, the variance will be large.

**Property F:**

$$Var(U) = E(U^2) - (EU)^2 \qquad (4.4)$$

The term $E(U^2)$ is again evaluated using (3.34).

Thus for example, again take $X$ to be the number of dots which come up when we roll a die. Then, from (4.4),

$$Var(X) = E(X^2) - (EX)^2 \qquad (4.5)$$

Let's find that first term (we already know the second is $3.5^2$). From (3.34),

$$E(X^2) = \sum_{i=1}^{6} i^2 \cdot \frac{1}{6} = \frac{91}{6} \qquad (4.6)$$

Thus $Var(X) = E(X^2) - (EX)^2 = \frac{91}{6} - 3.5^2 = 2.92$, as before. Remember, though, that (4.4) is a shortcut formula for finding the variance, not the *definition* of variance.

Below is the derivation of (4.4). Keep in mind that $EU$ is a constant.

$$
\begin{aligned}
Var(U) &= E[(U - EU)^2] & (4.7)\\
&= E[U^2 - 2EU \cdot U + (EU)^2] \ \text{(algebra)} & (4.8)\\
&= E(U^2) + E(-2EU \cdot U) + E[(EU)^2] \ (3.26) & (4.9)\\
&= E(U^2) - 2EU \cdot EU + (EU)^2 \ (3.27),(3.30) & (4.10)\\
&= E(U^2) - (EU)^2 & (4.11)
\end{aligned}
$$

An important behavior of variance is:

**Property G:**

$$Var(cU) = c^2 Var(U) \qquad (4.12)$$

for any random variable $U$ and constant $c$. It should make sense to you: If we multiply a random variable by 5, say, then its average squared distance to its mean should increase by a factor of 25.

Let's prove (4.12). Define $V = cU$. Then

$$
\begin{aligned}
Var(V) &= E[(V - EV)^2] \text{ (def.)} &&(4.13)\\
&= E\{[cU - E(cU)]^2\} \text{ (subst.)} &&(4.14)\\
&= E\{[cU - cEU]^2\} \text{ ((3.27))} &&(4.15)\\
&= E\{c^2[U - EU]^2\} \text{ (algebra)} &&(4.16)\\
&= c^2 E\{[U - EU]^2\} \text{ ((3.27))} &&(4.17)\\
&= c^2 Var(U) \text{ (def.)} &&(4.18)
\end{aligned}
$$

Shifting data over by a constant does not change the amount of variation in them:

**Property H:**
$$
Var(U + d) = Var(U) \qquad\qquad (4.19)
$$

for any constant d.

This example may put variance in perspective:

### Chemistry Examination

Say a chemistry professor tells her class that the mean score on the exam was 62.3, with a standard deviation of 11.4. But there is good news! She is going to add 10 points to everyone's score. What will happen to the mean and standard deviation?

From (3.26), with $V$ being the constant 10, we see that the mean on the exam will rise by 10. But (4.19) says that the variance or standard deviation won't change.

Intuitively, the variance of a constant is 0 — after all, it never varies! You can show this formally using (4.4):

$$
Var(c) = E(c^2) - [E(c)]^2 = c^2 - c^2 = 0 \qquad\qquad (4.20)
$$

As with expected value, we use variance as our main measure of dispersion for historical and mathematical reasons, not because it's the most meaningful measure. The squaring in the definition of variance produces some distortion, by exaggerating the importance of the larger differences. It would be more natural to use the *mean absolute deviation* (MAD), $E(|U - EU|)$ (and even better to use the median$(U)$ in place of $EU$ here). However, this is less tractable mathematically, so the statistical pioneers chose to use the mean squared difference, which lends itself to lots of powerful and beautiful math, in which the Pythagorean Theorem pops up in abstract vector spaces. (Sadly, this is beyond the scope of this book!)

**As with expected values, the properties of variance discussed here, are key to the entire remainder of this book. You should notice immediately when you are in a setting in which they are applicable. For instance, if you see the variance of the sum of two random variables, you should instinctively think of (4.33) right away, and check whether they are independent.**

## 4.1.2   Central Importance of the Concept of Variance

No one needs to be convinced that the mean is a fundamental descriptor of the nature of a random variable. But the variance is of central importance too, and will be used constantly throughout the remainder of this book.

The next section gives a quantitative look at our notion of variance as a measure of dispersion.

## 4.1.3   Intuition Regarding the Size of Var(X)

*A billion here, a billion there, pretty soon, you're talking real money —* attributed to the late Senator Everett Dirksen, replying to a statement that some federal budget item cost "only" a billion dollars

Recall that the variance of a random variable $X$ is supposed to be a measure of the dispersion of $X$, meaning the amount that $X$ varies from one instance (one line in our notebook) to the next. But if $Var(X)$ is, say, 2.5, is that a lot of variability or not? We will pursue this question here.

### 4.1.3.1   Chebychev's Inequality

This inequality states that for a random variable $X$ with mean $\mu$ and variance $\sigma^2$,

$$P(|X - \mu| \geq c\sigma) \leq \frac{1}{c^2} \tag{4.21}$$

In other words, $X$ strays more than, say, 3 standard deviations from its mean at most only 1/9 of the time. This gives some concrete meaning to the notion that variance/standard deviation are measures of variation.

Again, consider the example of a chemistry exam:

> **Chemistry Examination**
>
> The professor mentions that anyone scoring more than 1.5 standard deviations above mean earns an A grade, while those with scores under 2.1 standard deviations below the mean get an F. You wonder, out of 200 students in the class, how many got either A or F grades?
>
> Take $c = 2.1$ in Chebychev. It tells us that at most $1/2.1^2 = 0.23$ of the students were in that category, about 46 of them.

We'll prove the inequality in Section 4.6.1.

### 4.1.3.2   The Coefficient of Variation

Continuing our discussion of the magnitude of a variance, look at our remark following (4.21):

> In other words, $X$ does not often stray more than, say, 3 standard deviations from its mean. This gives some concrete meaning to the concept of variance/standard deviation.

Or, think of the price of, say, widgets. If the price hovers around a \$1 million, but the variation around that figure is only about a dollar, you'd say there is essentially no variation. But a variation of about a dollar in the price of an ice cream cone would considered more substantial.

These considerations suggest that any discussion of the size of $Var(X)$ should relate to the size of $E(X)$. Accordingly, one often looks at the *coefficient of variation*, defined to be the ratio of the standard deviation to the mean:

$$\text{coef. of var.} = \frac{\sqrt{Var(X)}}{EX} \tag{4.22}$$

This is a scale-free measure (e.g., inches divided by inches), and serves as a good way to judge whether a variance is large or not.

## 4.2   A Useful Fact

For a random variable $X$, consider the function

$$g(c) = E[(X - c)^2] \qquad (4.23)$$

Remember, the quantity $E[(X-c)^2]$ is a number, so $g(c)$ really is a function, mapping a real number $c$ to some real output.

So, we can ask the question, What value of $c$ minimizes $g(c)$? To answer that question, write:

$$g(c) = E[(X - c)^2] = E(X^2 - 2cX + c^2) = E(X^2) - 2cEX + c^2 \quad (4.24)$$

where we have used the various properties of expected value derived earlier.

To make this concrete, suppose we are guessing people's weights — without seeing them and without knowing anything about them at all. (This is a somewhat artificial question, but it will become highly practical in Chapter 15.) Since we know nothing at all about these people, we will make the same guess for each of them.

What should that guess-in-common be? Your first inclination would be to guess everyone to be the mean weight of the population. If that value in our target population is, say, 142.8 pounds, then we'll guess everyone to be that weight. Actually, that guess turns out to be optimal in a certain sense, as follows.

Say $X$ is a person's weight. It's a random variable, because these people are showing up at random from the population. Then $X - c$ is our prediction error. How well will do in our predictions? We can't measure that as

$$E(\text{error}) \qquad (4.25)$$

because a lot of the positive and negative errors would cancel out. A reasonable measure would be

$$E(|X - c|) \qquad (4.26)$$

However, due to tradition, we use

$$E[(X - c)^2]  \tag{4.27}$$

Now differentiate (4.24)with respect to $c$, and set the result to 0. Remembering that $E(X^2)$ and $EX$ are constants, we have

$$0 = -2EX + 2c  \tag{4.28}$$

In other words, the minimum value of $E[(X - c)^2]$ occurs at $c = EX$. Our intuition was right!

Moreover: Plugging $c = EX$ into (4.24) shows that the minimum value of g(c) is $E(X - EX)^2]$ , which is $Var(X)$!

In notebook terms, think of guessing many, many people, meaning many lines in the notebook, one per person. Then (4.27) is the long-run average squared error in our guesses, and we find that we minimize that by guessing everyone's weight to be the population mean weight.

## 4.3   Covariance

This is a topic we'll cover fully in Chapter 11 but at least introduce here.

A measure of the degree to which $U$ and $V$ vary together is their *covariance*,

$$Cov(U, V) = E[(U - EU)(V - EV)]  \tag{4.29}$$

Except for a divisor to be introduced later, this is essentially *correlation*. Suppose, for instance, $U$ is usually large (relative to its expectation) at the same time $V$ is small (relative to its expectation). Think of the price of some item, say the economists' favorite, widgets. Though many issues come into play, generally stores charging a higher price $U$ will sell fewer of them $V$, and vice versa.

In other words, the quantity $(U - EU)(V - EV)$ will usually be negative; either the first factor is positive and the second negative, or vice versa. That implies that (4.29) will likely be negative.

On the other hand, suppose we have $U$ as human height and $V$ as weight. These are usually large together or small together, so the covariance will be positive.

So, covariance is basically what is referred to as "correlation" in common speech. In notebook terms, think of the lines in the notebook for people who are taller than average, i.e., for whom $U - EU > 0$. Most such people are also heavier than average, i.e., $V - EV > 0$, so that $(U - EU)(V - EV) > 0$. On the other hand, shorter people also tend to be lighter, so most lines with shorter people will have $U - EU < 0$ and $V - EV < 0$ — but still $(U - EU)(V - EV) > 0$. In other words, the long-run average of the $(U - EU)(V - EV)$ column will be positive.

The point is that, if two variables are positively related, e.g., height and weight, their covariance should be positive. This is the intuition underlying defining covariance as in (4.29). Clearly the sign of a covariance is of interest, though we'll see that the magnitude matters a lot too.

Again, one can use the properties of $E()$ to show that

$$Cov(U, V) = E(UV) - EU \cdot EV \qquad (4.30)$$

This will be derived fully in Chapter 11, but think about how to derive it yourself. Just use our old mailing tubes, e.g., $E(X + Y) = EX + EY$, $E(cX)$ for a constant $c$, etc. Keep in mind that $EU$ and $EV$ are constants!

Also

$$Var(U + V) = Var(U) + Var(V) + 2\, Cov(U, V) \qquad (4.31)$$

and more generally,

$$Var(aU + bV) = a^2\, Var(U) + b^2\, Var(V) + 2ab\, Cov(U, V) \qquad (4.32)$$

for any constants $a$ and $b$.

If $U$ and $V$ are independent, then $Cov(U, V) = 0$. In that case,

$$Var(U + V) = Var(U) + Var(V) \qquad (4.33)$$

Generalizing (4.32), for constants $a_1, ..., a_k$ and random variables $X_1, ..., X_k$, form the new random variable $a_1X_1 + ... + a_kX_k$. Then

$$Var(a_1X_1 + ... + a_kX_k) = \sum_{i=1}^{k} a_i^2 Var(X_i) + 2 \sum_{1 \le i < j \le k} a_ia_j\, Cov(X_i, X_j)$$

$$(4.34)$$

If the $X_i$ are independent, then we have the special case

$$Var(a_1X_1 + ... + a_kX_k) = \sum_{i=1}^{k} a_i^2 Var(X_i) \qquad (4.35)$$

## 4.4   Indicator Random Variables, and Their Means and Variances

**Definition 7** *A random variable that has the value 1 or 0, according to whether a specified event occurs or not, is called an indicator random variable for that event.*

You'll often see later in this book that the notion of an indicator random variable is a very handy device in certain derivations. But for now, let's establish its properties in terms of mean and variance.

> **Handy facts:** Suppose $X$ is an indicator random variable for the event $A$. Let $p$ denote $P(A)$. Then

$$E(X) = p \qquad (4.36)$$

$$Var(X) = p(1 - p) \qquad (4.37)$$

These two facts are easily derived. In the first case we have, using our properties for expected value,

$$EX = 1 \cdot P(X = 1) + 0 \cdot P(X = 0) = P(X = 1) = P(A) = p \qquad (4.38)$$

The derivation for $Var(X)$ is similar (use (4.4)).

For example, say Coin A has probability 0.6 of heads, Coin B is fair, and Coin C has probability 0.2 of heads. I toss A once, getting $X$ heads, then toss B once, obtaining $Y$ heads, then toss C once, resulting in $Z$ heads. Let $W = X + Y + Z$, i.e., the total number of heads from the three tosses ($W$ ranges from 0 to 3). Let's find $P(W = 1)$ and $Var(W)$.

We first must use old methods:

$$
\begin{aligned}
P(W = 1) &= P(X = 1 \text{ and } Y = 0 \text{ and } Z = 0 \text{ or } ...) & (4.39) \\
&= 0.6 \cdot 0.5 \cdot 0.8 + 0.4 \cdot 0.5 \cdot 0.8 + 0.4 \cdot 0.5 \cdot 0.2 & (4.40) \\
&= 0.44 & (4.41)
\end{aligned}
$$

For $Var(W)$, let's use what we just learned about indicator random variables; each of $X$, $Y$ and $Z$ are such variables. $Var(W) = Var(X) + Var(Y) + Var(Z)$, by independence and (4.33). Since $X$ is an indicator random variable, $Var(X) = 0.6 \cdot 0.4$, etc. The answer is then

$$
0.6 \cdot 0.4 + 0.5 \cdot 0.5 + 0.2 \cdot 0.8 = 0.65 \tag{4.42}
$$

## 4.4.1 Example: Return Time for Library Books, Version I

Suppose at some public library, patrons return books exactly 7 days after borrowing them, never early or late. However, they are allowed to return their books to another branch, rather than the branch where they borrowed their books. In that situation, it takes 9 days for a book to return to its proper library, as opposed to the normal 7. Suppose 50% of patrons return their books to a "foreign" library. Find $Var(T)$, where $T$ is the time, either 7 or 9 days, for a book to come back to its proper location.

Note that

$$
T = 7 + 2I, \tag{4.43}
$$

where I is an indicator random variable for the event that the book is returned to a "foreign" branch. Then

$$
Var(T) = Var(7 + 2I) = 4Var(I) = 4 \cdot 0.5(1 - 0.5) = 1 \tag{4.44}
$$

## 4.4.2   Example: Return Time for Library Books, Version II

Now let's look at a somewhat broader model. Here we will assume that borrowers return books after 4, 5, 6 or 7 days, with probabilities 0.1, 0.2, 0.3, 0.4, respectively. As before, 50% of patrons return their books to a "foreign" branch, resulting in an extra 2-day delay before the book arrives back to its proper location. The library is open 7 days a week.

Suppose you wish to borrow a certain book, and inquire at the library near the close of business on Monday. Assume too that no one else is waiting for the book. You are told that it had been checked out the previous Thursday. Find the probability that you will need to wait until Wednesday evening to get the book. (You check every evening.)

Let $B$ denote the time needed for the book to arrive back at its home branch, and define $I$ as before. Then, as usual, translating English to math, we see that we are given that $B > 4$ (the book is not yet back, 4 days after it was checked out), and

$$
\begin{aligned}
P(B = 6 \mid B > 4) &= \frac{P(B = 6 \text{ and } B > 4)}{P(B > 4)} \\
&= \frac{P(B = 6)}{P(B > 4)} \\
&= \frac{P(B = 6 \text{ and } I = 0 \ \text{ or } \ B = 6 \ \text{ and } \ I = 1)}{1 - P(B = 4)} \\
&= \frac{0.5 \cdot 0.3 + 0.5 \cdot 0.1}{1 - 0.5 \cdot 0.1} \\
&= \frac{4}{19}
\end{aligned}
$$

The denominator in that third equality reflects the fact that borrowers always return a book after at least 4 days. In the numerator, as usual we used our "break big events down into small events" tip.

Here is a simulation check:

```
libsim <- function(nreps) {
    # patron return time
    prt <- sample(c(4,5,6,7),nreps,replace=TRUE,
        prob=c(0.1,0.2,0.3,0.4))
    # indicator for foreign branch
```

```
i <- sample(c(0,1),nreps,replace=TRUE)
b <- prt + 2*i
x <- cbind(prt,i,b)
# look only at the relevant notebook lines
bgt4 <- x[b > 4,]
# among those lines, what proportion have B = 6?
mean(bgt4[,3] == 6)
}
```

Note that in this simulation. all **nreps** values of $I$, $B$ and the patron return time are generated first. This uses more memory space (though not an issue in this small problem), but makes things easier to code, as we can exploit R's vector operations. Those not only are more convenient, but also faster running.

### 4.4.3 Example: Indicator Variables in a Committee Problem

A committee of four people is drawn at random from a set of six men and three women. Suppose we are concerned that there may be quite a gender imbalance in the membership of the committee. Toward that end, let $M$ and $W$ denote the numbers of men and women in our committee, and let the difference be $D = M - W$. Let's find $E(D)$, in two different ways.

$D$ has support consisting of the values 4-0, 3-1, 2-2 and 1-3, i.e., 4, 2, 0 and -2. So from (3.19)

$$ED = -2 \cdot P(D = -2) + 0 \cdot P(D = 0) + 2 \cdot P(D = 2) + 4 \cdot P(D = 4) \quad (4.45)$$

Now, using reasoning along the lines in Section 1.11, we have

$$P(D = -2) = P(M = 1 \text{ and } W = 3) = \frac{\binom{6}{1}\binom{3}{3}}{\binom{9}{4}} \quad (4.46)$$

After similar calculations for the other probabilities in (4.45), we find the $ED = \frac{4}{3}$.

Note what this means: If we were to perform this experiment many times, i.e., choose committees again and again, on average we would have a little more than one more man than women on the committee.

Now let's use our "mailing tubes" to derive $ED$ a different way:

$$
\begin{aligned}
ED &= E(M - W) & (4.47)\\
&= E[M - (4 - M)] & (4.48)\\
&= E(2M - 4) & (4.49)\\
&= 2EM - 4 \ \ \text{(from (3.28))} & (4.50)
\end{aligned}
$$

Now, let's find $EM$ by using indicator random variables. Let $G_i$ denote the indicator random variable for the event that the $i^{th}$ person we pick is male, $i = 1, 2, 3, 4$. Then

$$
M = G_1 + G_2 + G_3 + G_4 \tag{4.51}
$$

so

$$
\begin{aligned}
EM &= E(G_1 + G_2 + G_3 + G_4)\\
&= EG_1 + EG_2 + EG_3 + EG_4 \ \ [\text{ from (3.26)}]\\
&= P(G_1 = 1) + P(G_2 = 1) + P(G_3 = 1) + P(G_4 = 1) \ [\text{ fr. (4.36)}]
\end{aligned}
$$

Note carefully that the second equality here, which uses (3.26), is true in spite of the fact that the $G_i$ are not independent. Equation (3.26) does not require independence.

Another key point is that, due to symmetry, $P(G_i = 1)$ is the same for all i. Note that we did not write a *conditional* probability here! For instance, we are NOT talking about, say, $P(G_2 = 1 \mid G_1 = 1)$. Once again, think of the notebook view: **By definition**, $P(G_2 = 1)$ is the long-run proportion of the number of notebook lines in which $G_2 = 1$ — regardless of the value of $G_1$ in those lines.

Now, to see that $P(G_i = 1)$ is the same for all i, suppose the six men that are available for the committee are named Alex, Bo, Carlo, David, Eduardo and Frank. When we select our first person, any of these men has the same chance of being chosen, 1/9. *But that is also true for the second pick.* Think of a notebook, with a column named Second Pick. Don't peek at the First Pick column — it's not relevant to $P(G_2 = 1)$! In some lines, that column will say Alex, in some it will say Bo, and so on, and in some lines there will be women's names. But in that column, Bo will appear the same fraction of the time as Alex, due to symmetry, and that will be the same fraction as for, say, Alice, again 1/9. That probability is also 1/9 for Bo's being chosen first pick, third pick and fourth pick.

So,

$$P(G_1 = 1) = \frac{6}{9} = \frac{2}{3} \tag{4.52}$$

Thus

$$ED = 2 \cdot (4 \cdot \frac{2}{3}) - 4 = \frac{4}{3} \tag{4.53}$$

## 4.5  Skewness

We have seen the mean and variance os measures of central tendency and dispersion. In classical statistics, another common measure is *skewness*, measuring the degree to which a distribution is asymmetric about its mean. It is defined for a random variable $Z$ as

$$E\left[\left[\frac{Z - EZ}{\sqrt{Var(Z)}}\right]^3\right] \tag{4.54}$$

# 4.6  Mathematical Complements

## 4.6.1  Proof of Chebychev's Inequality

To prove (4.21), let's first state and prove Markov's Incquality: For any nonnegative random variable $Y$ and positive constant $d$,

$$P(Y \geq d) \leq \frac{EY}{d} \tag{4.55}$$

To prove (4.55), let $Z$ be the indicator random variable for the event $Y \geq d$.

Table 4.2: Illustration of Y and Z

| notebook line | Y | dZ | $Y \geq dZ$? |
|---|---|---|---|
| 1 | 0.36 | 0 | yes |
| 2 | 3.6 | 3 | yes |
| 3 | 2.6 | 0 | yes |

Now note that

$$Y \geq dZ \tag{4.56}$$

To see this, just think of a notebook, say with $d = 3$. Then the notebook might look like Table 4.2.

So

$$EY \geq d\ EZ \tag{4.57}$$

(Again think of the notebook. Due to (4.56), the long-run average in the $Y$ column will be $\geq$ the corresponding average for the $dZ$ column.)

The right-hand side of (4.56) is $d\ P(Y \geq d)$, so (4.55) follows.

Now to prove (4.21), define

$$Y = (X - \mu)^2 \tag{4.58}$$

and set $d = c^2\sigma^2$. Then (4.55) says

$$P[(X - \mu)^2 \geq c^2\sigma^2] \leq \frac{E[(X - \mu)^2]}{c^2\sigma^2} \tag{4.59}$$

The left-hand side is

$$P(|X - \mu| \geq c\sigma) \tag{4.60}$$

Meanwhile, the numerator the right-hand side is $\sigma^2$, by the very definition of variance. That gives us (4.21).

## 4.7 Exercises

**Mathematical problems:**

**1.** Consider the committee problem, Section 4.4.3. We of course chose without replacement there, but suppose we were to sample *with* replacement. What would be the new value of $E(D)$? Hint: This can be done without resorting to a lot of computation.

**2.** Suppose $Z$ is an indicator random variable with $P(Z = 1) = w$. Find the skewness of $Z$ (Section 4.5).

**3.** In the example in Section 4.4.3, find $Cov(M, W)$.

**4.** Suppose $X$ and $Y$ are indicator random variables with $P(X = 1) = P(Y = 1) = v$, and such that $P(X = Y = 1) = w$. Find $Var(X + Y)$ in terms of $v$ and $w$.

**5.** First show, citing the mailing tubes, that if $X$ and $Y$ are independent random variables, then $Var(X - Y) = Var(X) + Var(Y)$.

Now consider the bus ridership example. Intuitively, $L_1$ and $L_2$ are not independent. Confirm this by showing that the relation above does not hold with $X = L_2$ and $Y = L_0$. (Find the three variances analytically, and confirm via simulation.)

**6.** Consider the Preferential Attachment Model (Section 1.10.1), at the time immediately after $v4$ is added. Find the following, both mathematically and by simulation:

(a) Expected value and variance of the degree of $v_1$.

(b) Covariance between the degrees of $v_1$ and $v_2$.

**7.** Consider the board game example, Section 1.8, with the random variable B being the amount of bonus, 0,1,...,6 . Find $EB$ and $Var(B)$, both mathematically and via simulation.

**8.** Suppose $X$ and $Y$ are independent random variables, with $EX = 1$, $EY = 2$, $Var(X) = 3$ and $Var(Y) = 4$. Find $Var(XY)$. (The reader should make sure to supply the reasons for each step, citing equation numbers from the material above.)

**9.** Consider the board game example, Section 1.8. Let $X$ denote the number of turns needed to reach or pass 0. (Do not count a bonus roll as a separate turn.) Find $Var(X)$.

**10**. In the chemistry examination example, page 70, find an upper bound on the number of Fs.

**11**. Say we toss a coin 8 times, resulting in a H,T pattern, e.g., HHTHTHTT. Let $X$ denote the number of instances of HTH, e.g., two in HHTHTHTT. Find $EX$. Hint: Use indicator variables.

**Computational and data problems:**

**12**. Consider the broken rod example in Section 2.6. Find the variance of the minimum piece length, using simulation.

**13**. Using the simulation code developed in Problem 3, Chapter 2, find the variance of the degree of $v_1$ after $v_5$ joins the network.

# Chapter 5

# Discrete Parametric Distribution Families

There are a few famous probability models that are widely used in all kinds of applications. We introduce them in this chapter and Chapter 6.

## 5.1 Distributions

The idea of the *distribution* of a random variable is central to probability and statistics.

**Definition 8** *Let U be a discrete random variable. Then the distribution of U is simply the support of U, together with the associated probabilities.*

**Example:** Let $X$ denote the number of dots one gets in rolling a die. Then the values $X$ can take on are 1,2,3,4,5,6, each with probability 1/6. So

$$\text{distribution of } X = \{(1, \frac{1}{6}), (2, \frac{1}{6}), (3, \frac{1}{6}), (4, \frac{1}{6}), (5, \frac{1}{6}), (6, \frac{1}{6})\} \quad (5.1)$$

**Example:** Recall the ALOHA example. There $X_1$ took on the values 1 and 2, with probabilities 0.48 and 0.52, respectively (the case of 0 was impossible). So,

$$\text{distribution of } X_1 = \{(1, 0.48), (2, 0.52)\} \qquad (5.2)$$

**Example:** Recall our example in which $N$ is the number of tosses of a coin needed to get the first head. $N$ has support 1,2,3,..., the probabilities of which we found earlier to be 1/2, 1/4, 1/8,... So,

$$\text{distribution of } N = \{(1, \frac{1}{2}), (2, \frac{1}{4}), (3, \frac{1}{8}), ...\} \qquad (5.3)$$

We usually express this in functional notation:

**Definition 9** *The probability mass function (pmf) of a discrete random variable $V$, denoted $p_V$, is*

$$p_V(k) = P(V = k) \qquad (5.4)$$

*for any value $k$ in the support of $V$.*

(Please keep in mind the notation, as **it will be used extensively throughout the book.** It is customary to use the lower-case p, with a subscript consisting of the name of the random variable.)

Note that $p_V()$ is just a function, like any function (with integer domain) you've had in your previous math courses. For each input value, there is an output value.

## 5.1.1    Example: Toss Coin Until First Head

In (5.3),

$$p_N(k) = \frac{1}{2^k}, k = 1, 2, ... \qquad (5.5)$$

## 5.1.2 Example: Sum of Two Dice

In the dice example, in which $S = X + Y$,

$$
p_S(k) = \begin{cases} \frac{1}{36}, & k = 2 \\ \frac{2}{36}, & k = 3 \\ \frac{3}{36}, & k = 4 \\ \dots \\ \frac{1}{36}, & k = 12 \end{cases} \tag{5.6}
$$

It is important to note that there may not be some nice closed-form expression for $p_V$ like that of (5.5). There was no such form in (5.6), nor is there in our ALOHA example for $p_{X_1}$ and $p_{X_2}$.

## 5.1.3 Example: Watts-Strogatz Random Graph Model

Random graph models are used to analyze many types of link systems, such as power grids, social networks and even movie stars. We saw our first example model in Section 1.10.1; here is another, due to Duncan Watts and Steven Strogatz [42].

### 5.1.3.1 The Model

We have a graph of $n$ nodes, e.g., in which each node is a person).[1] Think of them as being linked in a circle — we're just talking about relations here, not physical locations — so we already have $n$ links. One can thus reach any node in the graph from any other, by following the links of the circle. (We'll assume all links are bidirectional.)

We now randomly add $k$ more links ($k$ is thus a parameter of the model), which will serve as "shortcuts." There are $\binom{n}{2} = n(n-1)/2$ possible links between nodes, but remember, we already have $n$ of those in the graph, so there are only $n(n-1)/2 - n = n^2/2 - 3n/2$ possibilities left. We'll be forming $k$ new links, chosen at random from those $n^2/2 - 3n/2$ possibilities.

Let $M$ denote the number of links attached to a particular node, which you may recall is known as the *degree* of a node. $M$ is a random variable (we

---

[1] The word *graph* here doesn't mean "graph" in the sense of a picture. Here we are using the computer science sense of the word, meaning a system of vertices and edges. It's also common to call those *nodes* and *links*.

are choosing the shortcut links randomly), so we can talk of its pmf, $p_M$, termed the *degree distribution* of the graph, which we'll calculate now.

Well, $p_M(r)$ is the probability that this node has $r$ links. Since the node already had 2 circle links before the shortcuts were constructed, $p_M(r)$ is the probability that $r - 2$ of the $k$ shortcuts attach to this node.

Other than the two neighboring links in the original circle and the "link" of a node to itself, there are n-3 possible shortcut links to attach to our given node. We're interested in the probability that $r - 2$ of them are chosen, and that $k - (r - 2)$ are chosen from the other possible links. Thus our probability is:

$$ p_M(r) = \frac{\binom{n-3}{r-2}\binom{n^2/2-3n/2-(n-3)}{k-(r-2)}}{\binom{n^2/2-3n/2}{k}} = \frac{\binom{n-3}{r-2}\binom{n^2/2-5n/2+3}{k-(r-2)}}{\binom{n^2/2-3n/2}{k}} \tag{5.7} $$

## 5.2   Parametric Families of Distributions

The notion of a *parametric family* of distributions is a key concept that will recur throughout the book.

Consider plotting the curves $g_{a,b}(x) = (x - a)^2 + b$. For each $a$ and $b$, we get a different parabola, as seen in the plot of three of the curves in Figure 5.1.

This is a family of curves, thus a family of functions. We say the numbers $a$ and $b$ are the *parameters* of the family. Note carefully that $x$ is not a parameter, but rather just an argument of each function. The point is that $a$ and $b$ are indexing the curves.

## 5.3   The Case of Importance to Us: Parameteric Families of pmfs

Probability mass functions are still functions.[2] Thus they too can come in parametric families, indexed by one or more parameters. We had an example in Section 5.1.3. Since we get a different function $p_M$ for each

---

[2]The domains of these functions are typically the integers, but that is irrelevant; a function is a function.

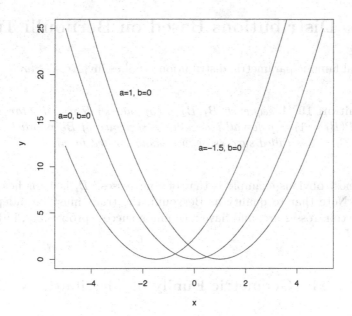

Figure 5.1: A parametric family of parabolas

different values of $k$ and $n$, that was a parametric family of pmfs, indexed by $k$ and $n$.

Some parametric families of pmfs have been found to be so useful over the years that they've been given names. We will discuss some of those families in this chapter. But remember, they are famous just because they have been found useful, i.e., that they fit real data well in various settings. **Do not jump to the conclusion that we always "must" use pmfs from some family.**

## 5.4   Distributions Based on Bernoulli Trials

Several famous parametric distribution families involve *Bernoulli trials*:

**Definition 10** *A sequence $B_1, B_2, \ldots$ of independent indicator variables with $P(B_i = 1) = p$ for all $i$ is called a sequence of Bernoulli trials. The event $B_i = 1$ is called success, with 0 being termed failure.*[3]

The most obvious example is that of coin tosses. $B_i$ is 1 for heads, 0 for tails. Note that to quality as Bernoulli, the trials must be independent, as the coin tosses are, and have a common success probability, in this case $p = 0.5$.

### 5.4.1   The Geometric Family of Distributions

Our first famous parametric family of pmfs involves the number of trials needed to obtain the first success.

Recall our example of tossing a coin until we get the first head, with $N$ denoting the number of tosses needed. In order for this to take $k$ tosses, we need $k - 1$ tails and then a head. Thus

$$p_N(k) = (1 - \frac{1}{2})^{k-1} \cdot \frac{1}{2}, \ k = 1, 2, \ldots \tag{5.8}$$

We say that $N$ has a geometric distribution with $p = 1/2$.

---

[3]These are merely labels, not meant to connote "good" and "bad."

We might call getting a head a "success," and refer to a tail as a "failure." Of course, these words don't mean anything; we simply refer to the outcome of interest (which of course we ourselves choose) as "success."

Define $M$ to be the number of rolls of a die needed until the number 5 shows up. Then

$$p_M(k) = \left(1 - \frac{1}{6}\right)^{k-1} \frac{1}{6}, \ k = 1, 2, ... \tag{5.9}$$

reflecting the fact that the event $M = k$ occurs if we get $k - 1$ non-5s and then a 5. Here "success" is getting a 5.

We say that $N$ has a geometric distribution with $p = 1/6$.

In general, suppose the random variable $W$ is defined to be the number of trials needed to get a success in a sequence of Bernoulli trials. Then

$$p_W(k) = (1 - p)^{k-1}p, \ k = 1, 2, ... \tag{5.10}$$

Note that there is a different distribution for each value of $p$, so we call this a *parametric family* of distributions, indexed by the parameter $p$. We say that $W$ is *geometrically distributed* with parameter $p$.[4]

It should make good intuitive sense to you that

$$E(W) = \frac{1}{p} \tag{5.11}$$

This is indeed true, which we will now derive. First we'll need some facts (which you should file mentally for future use as well):

**Properties of Geometric Series:**

(a) For any $t \neq 1$ and any nonnegative integers $r \leq s$,

$$\sum_{i=r}^{s} t^i = t^r \frac{1 - t^{s-r+1}}{1 - t} \tag{5.12}$$

This is easy to derive for the case $r = 0$, using mathematical induction. For the general case, just factor out $t^r$.

---

[4]Unfortunately, we have overloaded the letter $p$ here, using it to denote the probability mass function on the left side, and the unrelated parameter $p$, our success probability on the right side. It's not a problem as long as you are aware of it, though.

(b) For $|t| < 1$,

$$\sum_{i=0}^{\infty} t^i = \frac{1}{1-t} \tag{5.13}$$

To prove this, just take $r = 0$ and let $s \to \infty$ in (5.12).

(c) For $|t| < 1$,

$$\sum_{i=1}^{\infty} i t^{i-1} = \frac{1}{(1-t)^2} \tag{5.14}$$

This is derived by applying $\frac{d}{dt}$ to (5.13).[5]

Deriving (5.11) is then easy, using (3.19) and (5.14):

$$EW = \sum_{i=1}^{\infty} i(1-p)^{i-1}p \tag{5.15}$$

$$= p\sum_{i=1}^{\infty} i(1-p)^{i-1} \tag{5.16}$$

$$= p \cdot \frac{1}{[1-(1-p)]^2} \tag{5.17}$$

$$= \frac{1}{p} \tag{5.18}$$

Using similar computations, one can show that

$$Var(W) = \frac{1-p}{p^2} \tag{5.19}$$

We can also find a closed-form expression for the quantities $P(W \leq m)$, $m = 1, 2, \ldots$ This has a formal name, the *cumulative distribution function* (cdf), denoted $F_W(m)$, as will be seen later in Section 6.3. For any positive integer $m$ we have

---

[5]To be more careful, we should differentiate (5.12) and take limits.

$$
\begin{aligned}
F_W(m) &= P(W \le m) && (5.20) \\
&= 1 - P(W > m) && (5.21) \\
&= 1 - P(\text{the first m trials are all failures}) && (5.22) \\
&= 1 - (1-p)^m && (5.23)
\end{aligned}
$$

By the way, if we were to think of an experiment involving a geometric distribution in terms of our notebook idea, the notebook would have an infinite number of columns, one for each Bernoulli trial $B_i$. Within each row of the notebook, the $B_i$ entries would be 0 until the first 1, then NA ("not applicable") after that.

### 5.4.1.1   R Functions

You can simulate geometrically distributed variables via R's **rgeom()** function. Its first argument specifies the number of such random variables you wish to generate, and the second is the success probability $p$.

For example, if you run

```
> y <- rgeom(2,0.5)
```

then it's simulating tossing a coin until you get a head (**y[1]** tosses needed) and then tossing the coin until a head again (**y[2]** tosses). Of course, you could simulate on your own, say using **sample()** and **while()**, but R makes it convenient for you.

Here's the full set of functions for a geometrically distributed random variable $X$ with success probability $p$:

- **dgeom(i,p)**, to find $P(X = i)$ (pmf)

- **pgeom(i,p)**, to find $P(X \le i)$ (cdf)

- **qgeom(q,p)**, to find $c$ such that $P(X \le c) = q$ (inverse cdf)

- **rgeom(n,p)**, to generate **n** variates from this geometric distribution

**Important note:** Though our definition here is fairly standard, some books define geometric distributions slightly differently, as the number of failures before the first success, rather than the number of trials to the first success. The same is true for software—both R and Python define it this way. Thus for example in calling **dgeom()**, in the context of our definition, use **i-1** rather than **i** in the argument.

For example, here is $P(N = 3)$ for a geometric distribution under our defintion, with $p = 0.4$:

```
> dgeom(2,0.4)
[1] 0.144
> # check
> (1-0.4)^(3-1) * 0.4
[1] 0.144
```

Note that this also means one must *add* 1 to the result of **rgeom()**.


### 5.4.1.2   Example: A Parking Space Problem

Suppose there are 10 parking spaces per block on a certain street. You turn onto the street at the start of one block, and your destination is at the start of the next block. You take the first parking space you encounter. Let $D$ denote the distance of the parking place you find from your destination, measured in parking spaces. In this simple model, suppose each space is open with probability 0.15, with the spaces being independent. Find $ED$.

To solve this problem, you might at first think that D follows a geometric distribution. **But don't jump to conclusions!** Actually this is not the case; D is a somewhat complicated distance. But clearly $D$ is a function of $N$, where the latter denotes the number of parking spaces you see until you find an empty one—and $N$ *is* geometrically distributed.

As noted, $D$ is a function of $N$:

$$D = \begin{cases} 11 - N, & N \leq 10 \\ N - 11, & N > 10 \end{cases} \tag{5.24}$$

Since $D$ is a function of $N$, we can use (3.34) with $g(t)$ as in (5.24):

$$ED = \sum_{i=1}^{10}(11 - i)(1 - 0.15)^{i-1}0.15 + \sum_{i=11}^{\infty}(i - 11)0.85^{i-1}0.15 \tag{5.25}$$

This can now be evaluated using the properties of geometric series presented above.

Alternatively, here's how we could find the result by simulation:

```
parksim <- function(nreps) {
    # do the experiment nreps times,
    # recording the values of N
    nvals <- rgeom(nreps,0.15) + 1
    # now find the values of D
    dvals <- abs(nvals - 11)
    # return ED
    mean(dvals)
}
```

Note the vectorized addition and recycling (Section 2.1.2) in the line

```
nvals <- rgeom(nreps,0.15) + 1
```

The call to **abs()** is another instance of R's vectorization.

Let's find some more, first $p_N(3)$:

$$p_N(3) = P(N = 3) = (1 - 0.15)^{3-1}0.15 \tag{5.26}$$

Next, find $P(D = 1)$:

$$
\begin{aligned}
P(D = 1) &= P(N = 10 \text{ or } N = 12) \tag{5.27} \\
&= (1 - 0.15)^{10-1}0.15 + (1 - 0.15)^{12-1}0.15 \tag{5.28}
\end{aligned}
$$

Say Joe is the one looking for the parking place. Paul is watching from a side street at the end of the first block (the one before the destination), and Martha is watching from an alley situated right after the sixth parking space in the second block. Martha calls Paul and reports that Joe never went past the alley, and Paul replies that he did see Joe go past the first block. They are interested in the probability that Joe parked in the second space in the second block. In mathematical terms, what probability is that? Make sure you understand that it is $P(N = 12 \mid N > 10 \text{ and } N \leq 16)$. It can be evaluated as above.

Or consider a different question: Good news! I found a parking place just one space away from the destination. Find the probability that I am parked in the same block as the destination.

$$P(N = 12 \mid N = 10 \text{ or } N = 12) = \frac{P(N = 12)}{P(N = 10 \text{ or } N = 12)}$$

$$= \frac{(1 - 0.15)^{11} \, 0.15}{(1 - 0.15)^9 \, 0.15 + (1 - 0.15)^{11} \, 0.15}$$

## 5.4.2   The Binomial Family of Distributions

A geometric distribution arises when we have Bernoulli trials with parameter $p$, with a variable number of trials ($N$) but a fixed number of successes (1). A *binomial distribution* arises when we have the opposite situation — a fixed number of Bernoulli trials ($n$) but a variable number of successes (say $X$).[6]

For example, say we toss a coin five times, and let $X$ be the number of heads we get. We say that $X$ is binomially distributed with parameters $n = 5$ and $p = 1/2$. Let's find $P(X = 2)$. There are many orders in which that could occur, such as HHTTT, TTHHT, HTTHT and so on. Each order has probability $0.5^2(1 - 0.5)^3$, and there are $\binom{5}{2}$ orders. Thus

$$P(X = 2) = \binom{5}{2}0.5^2(1 - 0.5)^3 = \binom{5}{2}/32 = 5/16 \qquad (5.29)$$

For general $n$ and $p$,

$$p_X(k) = P(X = k) = \binom{n}{k}p^k(1 - p)^{n-k} \qquad (5.30)$$

So again we have a parametric family of distributions, in this case a family having two parameters, $n$ and $p$.

Let's write $X$ as a sum of those 0-1 Bernoulli variables we used in the discussion of the geometric distribution above:

$$X = \sum_{i=1}^{n} B_i \qquad (5.31)$$

---

[6]Note again the custom of using capital letters for random variables, and lower-case letters for constants.

where $B_i$ is 1 or 0, depending on whether there is success on the $i^{th}$ trial or not. Note again that the $B_i$ are indicator random variables (Section 4.4), so

$$EB_i = p \qquad (5.32)$$

and

$$Var(B_i) = p(1-p) \qquad (5.33)$$

Then the reader should use our earlier properties of E() and Var() in Sections 3.5 and 4.1 to fill in the details in the following derivations of the expected value and variance of a binomial random variable:

$$EX = E(B_1 + ... + B_n) = EB_1 + ... + EB_n = np \qquad (5.34)$$

and from (4.33),

$$Var(X) = Var(B_1 + ... + B_n) = Var(B_1) + ... + Var(B_n) = np(1-p) \qquad (5.35)$$

Again, (5.34) should make good intuitive sense to you.

### 5.4.2.1 R Functions

Relevant functions for a binomially distributed random variable $X$ for $k$ trials and with success probability $p$ are:

- **dbinom(i,k,p)**, to find $P(X = i)$
- **pbinom(i,k,p)**, to find $P(X \leq i)$
- **qbinom(q,k,p)**, to find $c$ such that $P(X \leq c) = q$
- **rbinom(n,k,p)**, to generate $n$ independent values of $X$

Our definition above of **qbinom()** is not quite tight, though. Consider a random variable $X$ which has a binomial distribution with $n = 2$ and $p = 0.5$ Then

$$F_X(0) = 0.25, \quad F_X(1) = 0.75 \qquad (5.36)$$

So if $q$ is, say, 0.33, there is no c such that $P(X \le c) = q$. For that reason, the actual definition of **qbinom()** is the smallest $c$ satisfying $P(X \le c) \ge q$. (Of course, this was also an issue for **qgeom()**.)

### 5.4.2.2　Example: Parking Space Model

Recall Section 5.4.1.2. Let's find the probability that there are three open spaces in the first block.

Let $M$ denote the number of open spaces in the first block. This fits the definition of binomially-distributed random variables: We have a fixed number (10) of independent Bernoulli trials, and we are interested in the number of successes. So, for instance,

$$p_M(3) = \binom{10}{3} 0.15^3 (1 - 0.15)^{10-3} \tag{5.37}$$

## 5.4.3　The Negative Binomial Family of Distributions

Recall that a typical example of the geometric distribution family (Section 5.4.1) arises as $N$, the number of tosses of a coin needed to get our first head. Now generalize that, with $N$ now being the number of tosses needed to get our $r^{th}$ head, where $r$ is a fixed value. Let's find $P(N = k), k = r, r + 1, \ldots$ For concreteness, look at the case $r = 3, k = 5$. In other words, we are finding the probability that it will take us 5 tosses to accumulate 3 heads.

First note the equivalence of two events:

$$\{N = 5\} = \{2 \text{ heads in the first 4 tosses and head on the } 5^{th} \text{ toss}\} \tag{5.38}$$

That event described before the "and" corresponds to a binomial probability:

$$P(2 \text{ heads in the first 4 tosses}) = \binom{4}{2} \left(\frac{1}{2}\right)^4 \tag{5.39}$$

Since the probability of a head on the $k^{th}$ toss is 1/2 and the tosses are independent, we find that

$$P(N = 5) = \binom{4}{2}\left(\frac{1}{2}\right)^5 = \frac{3}{16} \qquad (5.40)$$

The negative binomial distribution family, indexed by parameters $r$ and $p$, corresponds to random variables that count the number of independent trials with success probability $p$ needed until we get $r$ successes. The pmf is

$$p_N(k) = P(N = k) = \binom{k-1}{r-1}(1-p)^{k-r}p^r, k = r, r+1, \ldots \qquad (5.41)$$

We can write

$$N = G_1 + \ldots + G_r \qquad (5.42)$$

where $G_i$ is the number of tosses between the success numbers $i - 1$ and $i$. But each $G_i$ has a geometric distribution! Since the mean of that distribution is $1/p$, we have that

$$E(N) = r \cdot \frac{1}{p} \qquad (5.43)$$

In fact, those $r$ geometric variables are also independent, so we know the variance of $N$ is the sum of their variances:

$$Var(N) = r \cdot \frac{1-p}{p^2} \qquad (5.44)$$

### 5.4.3.1   R Functions

Relevant functions for a negative binomial distributed random variable $X$ with success parameter $p$ are:

- **dnbinom(i,size=1,prob=p)**, to find $P(X = i)$
- **pnbinom(i,size=1,prob=p)**, to find $P(X <= i)$
- **qnbinom(q,sixe=1,prob=p)**, to find $c$ such that $P(X \le c) = q$

- **rnbinom(n,size=1,prob=p)**, to generate **n** independent values of $X$

Here **size** is our $r$. Note, though, that as with the **geom()** family, R defines the distribution in terms of number of failures. So, in **dbinom()**, the argument **i** is the number of failures, and **i** + **r** is our $X$.

### 5.4.3.2   Example: Backup Batteries

A machine contains one active battery and two spares. Each battery has a 0.1 chance of failure each month. Let $L$ denote the lifetime of the machine, i.e., the time in months until the third battery failure. Find $P(L = 12)$.

The number of months until the third failure has a negative binomial distribution, with $r = 3$ and $p = 0.1$. Thus the answer is obtained by (5.41), with $k = 12$:

$$P(L = 12) = \binom{11}{2}(1 - 0.1)^9 0.1^3 \tag{5.45}$$

## 5.5   Two Major Non-Bernoulli Models

The two distribution families in this section are prominent because they have been found empirically to fit well in many applications. This is in contrast to the geometric, binomial and negative binomial families, in the sense that in those cases there were qualitative descriptions of the settings in which such distributions arise. Geometrically distributed random variables, for example occur as the number of Bernoulli trials needed to get the first success, so the model comes directly from the structure of the process generating the data.

By contrast, the Poisson distribution family below is merely something that people have found to be a reasonably accurate model of actual data in many cases. We might be interested, say, in the number of disk drive failures in periods of a specified length of time. If we have data on this, we might graph it, and if it looks like the pmf form below, then we might adopt it as our model.[7]

---

[7]The Poisson family does also have some theoretical interest (Section 6.8.2) as well.

## 5.5.1   The Poisson Family of Distributions

The family of *Poisson Distributions* has pmf form

$$P(X = k) = \frac{e^{-\lambda}\lambda^k}{k!}, k = 0, 1, 2, ... \tag{5.46}$$

It turns out that

$$EX = \lambda \tag{5.47}$$

$$Var(X) = \lambda \tag{5.48}$$

The derivations of these facts are similar to those for the geometric family in Section 5.4.1. One starts with the Maclaurin series expansion for $e^t$:

$$e^t = \sum_{i=0}^{\infty} \frac{t^i}{i!} \tag{5.49}$$

and finds its derivative with respect to $t$, and so on. The details are left to the reader.

The Poisson family is very often used to model count data. For example, if you go to a certain bank every day and count the number of customers who arrive between 11:00 and 11:15 a.m., you will probably find that that distribution is well approximated by a Poisson distribution for some $\lambda$.

There is a lot more to the Poisson story than we see in this short section. We'll return to this distribution family in Section 6.8.2.

### 5.5.1.1   R Functions

Relevant functions for a Poisson distributed random variable X with parameter $\lambda$ are:

- **dpois(i,lambda)**, to find $P(X = i)$
- **ppois(i,lambda)**, to find $P(X \leq i)$
- **qpois(q,lambda)**, to find $c$ such that $P(X \leq c) = q$
- **rpois(n,lambda)**, to generate **n** independent values of $X$

### 5.5.1.2   Example: Broken Rod

Recall the example of a broken glass rod in Section 2.6. Suppose now that the number of breaks is random, not just the break points. A reasonable model to try would be Poisson. However, the latter's support starts at 0, and we cannot have 0 pieces, so we need to model the number of pieces minus 1 (the number of break points) as Poisson.

Suppose we wish to find the expected value of the shortest piece, via simulation. The code is similar to that in Section 2.6, but we must first generate the number of break points:

```
minpiecepois <- function(lambda) {
   nbreaks <- rpois(1,lambda) + 1
   breakpts <- sort(runif(nbreaks))
   lengths <- diff(c(0,breakpts,1))
   min(lengths)
}

bkrodpois <- function(nreps,lambda,q) {
   minpieces <-
      replicate(nreps,minpiecepois(lambda))
   mean(minpieces < q)
}

> bkrodpois(10000,5,0.02)
[1] 0.4655
```

Note that in each call to **minpiecepois()**, there will be a different number of breakpoints.

## 5.5.2   The Power Law Family of Distributions

This family has attracted quite a bit of attention in recent years, due to its use in random graph models.

### 5.5.2.1   The Model

Here

$$p_X(k) = ck^{-\gamma}, \ k = 1, 2, 3, \dots \tag{5.50}$$

It is required that $\gamma > 1$, as otherwise the sum of probabilities will be infinite. For $\gamma$ satisfying that condition, the value $c$ can be determined by noting that the sum is 1.0:

$$1.0 = \sum_{k=1}^{\infty} ck^{-\gamma} \approx c \int_{1}^{\infty} k^{-\gamma}\, dk = c/(\gamma - 1) \qquad (5.51)$$

so $c \approx \gamma - 1$.

Here again we have a parametric family of distributions, indexed by the parameter $\gamma$.

The power law family is an old-fashioned model (an old-fashioned term for *distribution* is *law*), but there has been a resurgence of interest in it in recent years. Analysts have found that many types of social networks in the real world exhibit approximately power law behavior in their degree distributions.

For instance, in a famous study [2] of degree distribution on the Web (a directed graph, with incoming links being the ones of interest here) it was found that the number of links leading to a Web page has an approximate power law distribution with $\gamma = 2.1$. The number of links leading out of a Web page was also found to be approximately power-law distributed, with $\gamma = 2.7$.

In addition, some of the theoretical models, such as the Preferential Attachment Model (Section 1.10.1), can be shown that after many generations, the degree distribution has form (5.50).

Much of the interest in power laws stems from their *fat tails*, a term meaning that values far from the mean are more likely under a power law than they would be under a normal distribution (the famous "bell-shaped curve," Section 6.7.2) with the same mean and standard deviation. In recent popular literature, values far from the mean have often been called *black swans*. The financial crash of 2008, for example, is blamed by some on *quants* (people who develop probabilistic models for guiding investment) underestimating the probabilities of values far from the mean.

Some examples of real data that are, or are not, fit well by power law models are given in [11]. A variant of the power law model is the *power law with exponential cutoff*, which essentially consists of a blend of the power law and a geometric distribution. Here

$$p_X(k) = ck^{-\gamma}q^k \qquad (5.52)$$

This now is a two-parameter family, the parameters being $\gamma$ and $q$. Again $c$ is determined by the fact that the pmf sums to 1.0.

This model is said to work better than a pure power law for some types of data. Note, though, that this version does not really have the fat tail property, as the tail decays exponentially now.

The interested reader will find further information in [11].

## 5.5.3 Fitting the Poisson and Power Law Models to Data

It was stated above that the popularity of the Poisson and power law models stems from the fact that they often fit real data well. How is that fit obtained, and how does one assess it?

Note that a dataset is treated as a sample from a larger source, idealized as a *population*. We'll cover this in detail in Chapter 7, but for now the point is that we must estimate population quantities from the data. We'll denote the data by $X_1, X_2, ..., X_n$.

### 5.5.3.1 Poisson Model

The Poisson family has a single parameter, $\lambda$, which happens to be the mean of the distribution. Given a dataset, we can find the average of the $X_i$, denoted $\overline{X}$, and take that as our estimate of $\lambda$. For $j = 0, 1, 2, ...$, our estimate for $P(X = j)$ under the Poisson model would then be

$$\frac{e^{\overline{X}}\overline{X}^j}{j!} \tag{5.53}$$

Without making the Poisson assumption, our estimate of $P(X = j)$ will be

$$\frac{\text{number of } X_i = j}{n} \tag{5.54}$$

We can then compare compare the two sets of estimates to assess the value of the Poisson model.

### 5.5.3.2    Straight-Line Graphical Test for the Power Law

Taking logarithms of both sides of (5.50), we have

$$\log p_X(k) = \log c - \gamma \log k \tag{5.55}$$

In other words, the graph of $\log p_X(k)$ against $\log k$ is a straight line with slope $-\gamma$. So if our data displays such a graph approximately, it would suggest that the power law is a good model for the quantity being measured. Moreover, the slope would provide us the estimated $\gamma$.

A key word above, though, is *approximately*. We are only working with data, not the population it comes from. We do not know the values of $p_X(k)$, and must estimate them from the data as the corresponding sample proportions. Hence we do not expect the data to exactly follow a straight line, merely that they follow a linear trend.

### 5.5.3.3    Example: DNC E-mail Data

The data are for a random graph of people, with a link between two people symbolizing that they were corecipients in at least one e-mail message under study.[8] Let's find their degree distribution, and see whether it seems to follow a power law.

Each row in the dataset has three columns, in the format

```
recipientA    recipientB    nmsgs
```

where the first two fields are the recipient IDs and the last is the number of messages. We'll treat this as a random graph, as in Section 5.1.3, with the above row considered as a link between the two nodes. We'll not use the third column.

Like many datasets, this one has problems. The description says it is an undirected graph, meaning that a link between recipients 1874 and 999, say, is considered the same as one between 999 and 1874. However, the dataset does also have a record for the latter (and with a different message count). Since this is just an illustrative example, we'll just take the records with the **recipientA** value smaller than **recipientB**, which turns out to cut the dataset approximately in half.

Here is the code:

---

[8]http://konect.uni-koblenz.de/networks/dnc-corecipient

```
recip1 <- recip[recip$V1 < recip$V2,]
degs <- tapply(recip1$V1,recip1$V1,length)
dtab <- table(degs)
plot(log(as.numeric(names(dtab))),log(dtab))
```

R's **tapply()** function is quite handy in many situations. There is a full presentation on this in Section 7.12.1.3, but the explanation of the above call is simply this: **tapply()** will form groups of values in the first argument, according to the set of unique values in the second, and then call **length()** on each group. The result is that **degs** contains the degree for each recipient in our dataset.

Here's what **degs** looks like:

```
> dtab
degs
    1    2    3    4    5    6    7    8    9   10   11   12   13
  167   60   38   39   22   17   12   15   19    9    6    7    9
 . . .
```

There were 167 recipients with one link, 60 recipients with 2 links and so on. There are 552 distinct recipients:

```
> sum(dtab)
[1] 552
```

The values 1, 2, 3 etc. are $k$ in (5.55), while 167/552, 60/552 and so on are our estimates of $p_X(k)$ as in (5.54). Thus their logs are estimates of the left-hand side of (5.55).[9]

The right side will be $\log 1$, $\log 2$ etc. The latter (without the logs) are the names of the entries in **degs**, which we convert to numbers using **as.numeric()**.[10]

The result, shown in Figure 5.2, does seem to show a linear trend. The trend is rather linear, though tailing off somewhat at the end, which is common for this kind of data.

We can fit a straight line and determine its slope using R's **lm()** function.[11] This function will be the star of the show in Chapter 15, so we will postpone

---

[9]In our plot code, we don't bother with dividing by 552, since we are just interested in viewing the trend in the data.

[10]Note that not all values of $k$ appear in the data; for instance, 33 and 45 are missing.

[11]In this case, the estimated slope turns out to be slightly smaller than 1.0, a seeming violation of the requirement $\gamma > 1.0$ for the power law family. But again, keep in mind that the value obtained from the data is just a sample estimate.

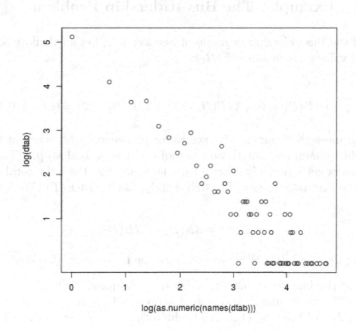

Figure 5.2: Log of degree distribution

details until then.

## 5.6    Further Examples

A bit more practice before moving on the continuous random variables in the next chapter.

### 5.6.1    Example: The Bus Ridership Problem

Recall the bus ridership example of Section 1.1. Let's calculate some expected values, for instance $E(B_1)$:

$$E(B_1) = 0 \cdot P(B_1 = 0) + 1 \cdot P(B_1 = 1) + 2 \cdot P(B_1 = 2) = 0.4 + 2 \cdot 0.1 \quad (5.56)$$

Now suppose the company charges \$3 for passengers who board at the first stop, but charges \$2 for those who join at the second stop. (The latter passengers get a possibly shorter ride, thus pay less.) So, the total revenue from the first two stops is $T = 3B_1 + 2B_2$. Let's find $E(T)$. We write

$$E(T) = 3E(B_1) + 2E(B_2) \quad (5.57)$$

making use of (3.28). We then compute the terms as in (5.56).

Suppose the bus driver has the habit of exclaiming, "What? No new passengers?!" every time he comes to a stop at which $B_i = 0$. Let $N$ denote the number of the stop (1,2,...) at which this first occurs. Find $P(N = 3)$:

$N$ has a geometric distribution, with $p$ equal to the probability that there are 0 new passengers at a stop, i.e., 0.5. Thus $p_N(3) = (1 - 0.5)^2 0.5$, by (5.10).

Also, let $S$ denote the number of stops, out of the first 6, at which 2 new passengers board. For example, S would be 3 if $B_1 = 2$, $B_2 = 2$, $B_3 = 0$, $B_4 = 1$, $B_5 = 0$, and $B_6 = 2$. Find $p_S(4)$:

$S$ has a binomial distribution, with $n = 6$ and $p =$ probability of 2 new passengers at a stop $= 0.1$. Then

$$p_S(4) = \binom{6}{4} 0.1^4 (1 - 0.1)^{6-4} \quad (5.58)$$

By the way, we can exploit our knowledge of binomial distributions to simplify the simulation code in Section 2.4. The lines

```
for (k in 1:passengers)
    if (runif(1) < 0.2)
        passengers <- passengers - 1
```

simulate finding the number of passengers who alight at that stop. But that number is binomially distributed, so the above code can be compactified (and speeded up in execution) as

```
passengers <-
    passengers - rbinom(1,passengers,0.2)
```

## 5.6.2 Example: Analysis of Social Networks

Let's continue our earlier discussion from Section 5.1.3.

One of the earliest—and now the simplest—models of social networks is due to Erdös and Renyi [13]. Say we have $n$ people (or $n$ Web sites, etc.), with $\binom{n}{2}$ potential links between pairs. (We are assuming an undirected graph here.) In this model, each potential link is an actual link with probability $p$, and a nonlink with probability $1 - p$, with all the potential links being independent.

Recall the notion of degree distribution from Section 5.1.3. Clearly the degree distribution $D_i$ here for a single node $i$ is binomial with parameters $n - 1$ and $p$.

But consider $k$ nodes, say 1 through $k$, among the $n$ total nodes, and let $T$ denote the number of links involving these nodes. Let's find the distribution of $T$. That distribution is again binomial, but the number of trials must be carefully calculated. We cannot simply say that, since each of the $k$ vertices can have as many as $n - 1$ links, there are $k(n-1)$ potential links, because there is overlap; two nodes among the $k$ have a potential link with each other, but we can't count it twice. So, let's reason this out.

Say $n = 9$ and $k = 4$. Among the four special nodes, there are $\binom{4}{2} = 6$ potential links, each on or off with probability $p$, independently. Also each of the four special nodes has $9 - 4 = 5$ potential links with the "outside world," i.e., the five non-special nodes. So there are $4 \times 5 = 20$ potential links here, for a total of 26.

So, the distribution of $T$ is binomial with

$$k(n-k) + \binom{k}{2} \tag{5.59}$$

trials and success probability $p$.

## 5.7   Computational Complements

### 5.7.1   Graphics and Visualization in R

R excels at graphics, offering a rich set of capabilities, from beginning to advanced. In addition to the extensive capability in base R, extensive graphics packages are available, such as **ggplot2** [44] and **lattice** [38]. A number of other books are available, including the definitive [35].

Here is the base-R code used to generate Figure 5.1:

```
prb <- function(x) x^2
prba <- function(x) (x-1)^2
prbb <- function(x) (x+1.5)^2
plot(curve(prb,-5,5),type='l',xlab='x',ylab='y')
lines(curve(prba,-5,5,add=TRUE),type='l')
lines(curve(prbb,-5,5,add=TRUE),type='l')
text(-2.3,18,'a=1, b=0')
text(1,12,'a=-1.5, b=0')
text(-4.65,15.5,'a=0, b=0')
```

Here are the main points:

- We defined functions for the three parabolas.

- We called **curve()**, which generates the points to be plotted for the given curve.

- We called **plot()**, with **type** $=$ 'l' signifying that we want a line rather than discrete points, and with the designated axis labels.

- We called **lines()**, which adds lines to an existing plot. The argument **add** informed **curve()** of this too.

- Finally, we called **text()** to add labels at the specified (X,Y) coordinates in the plot.

One common operation involves saving an R graph that is currently displayed on the screen to a file. Here is a function for this, which I include in my R startup file, **.Rprofile**, in my home directory:

```
pr2file <- function (filename)
{
    origdev <- dev.cur()
    parts <- strsplit(filename, ".", fixed = TRUE)
    nparts <- length(parts[[1]])
    suff <- parts[[1]][nparts]
    if (suff == "pdf") {
        pdf(filename)
    }
    else if (suff == "png") {
        png(filename)
    }
    else jpeg(filename)
    devnum <- dev.cur()
    dev.set(origdev)
    dev.copy(which = devnum)
    dev.set(devnum)
    dev.off()
    dev.set(origdev)
}
```

The code, which I won't go into here, mostly involves manipulation of various R graphics devices. I've set it up so that you can save to a file of type either PDF, PNG or JPEG, implied by the file name you give.

## 5.8 Exercises

**Mathematical problems:**

**1.** In the example of random groups of students, Section 1.11.2, let $X$ be the number of CS students chosen. Find $p_X()$.

**2.** In the example on lottery tickets, Section 1.11.3, let $X$ be the number of even-numbered tickets chosen. Find $p_X()$.

**3.** Consider the parking space example, Section 5.4.1.2. Find $P(D = 3)$ analytically, then by simulation, modifying the code in that section. Try to make your code loop-free.

**4.** Suppose $X_1$, $X_2$, ... are independent indicator variables but with different success probabilities; define $p_i = P(X_i = 1)$. Define $Y_n = X_1 + ... + X_n$. Find $EY_n$ and $Var(Y_n)$ in terms of the $p_i$.

**5.** For a discrete random variable $X$, its *hazard function* is defined as

$$h_X(k) = P(X = k + 1 \mid X > k) = \frac{p_X(k)}{1 - F_X(k)} \qquad (5.60)$$

The idea here is as follows: Say $X$ is battery lifetime in months. Then for instance $h_X(32)$ is the conditional probability that the battery will fail in the next month, given that it has lasted 32 months so far. The notion is widely used in medicine, insurance, device reliability and so on (though more commonly for continuous random variables than discrete ones). Show that for a geometrically distributed random variable, its hazard function is constant. We say that geometric random variables are *memoryless*: It doesn't matter how long some process has been going; the probability is the same that it will end in the next time epoch, as if it doesn't "remember" how long it has lasted so far.

**6.** In the Watts-Strogatz model, Section 5.1.3, find the probability that a specified node connects to exactly one shortcut.

**7.** Consider a geometrically distributed random variable $W$ with parameter $p$. Find a closed-form expression for $P(W$ is an even number$)$.

**Computational and data problems:**

**8.** In Section 3.5.2, it is stated that the probability of obtaining exactly 500 heads in 1000 tosses of a coin is about 0.025. Verify this.

**9.** Note the term *hazard function* in Problem 5. Write code to compute and plot the hazard function for a binomial distribution with $n = 10$ and $p = 0.4$. Also do this for a Poisson distribution with $\lambda = 3.5$.

**10.** Here you will develop "d,p,q,r" functions for a certain distribution family, in the sense of Sections 5.4.1.1, 5.4.2.1 and so on

We'll call the family "accum" for "accumulate." The setting is that of repeatedly rolling a pair of dice. The random variable $X$ is the number of rolls needed to achieve an accumulated total of at least $k$ dots. So for instance the support of $X$ ranges from ceiling($k/12$) to ceiling($k/2$). This is a one-parameter family.

Write functions **daccum()**, **paccum()** and so on. Try not to use simulation for the 'd' and 'p' cases; if you are comfortable with recursion, this may be the best approach. For the 'q' case, keep in mind the comment preceding (5.36).

**11.** Investigate how *tight* Chebychev's Inequality (Section 4.21) is, meaning how close the upper bound it provides is to the actual quantity. Specifically, say we roll a die 12 times, with $X$ denoting the number of rolls that yield a 5. Find the exact value of $P(X = 1$ or $X = 3)$, then compare it to the upper bound from Chebychev.

# Chapter 6

# Continuous Probability Models

There are other types of random variables besides the discrete ones we studied in Chapter 3. This chapter will cover another major class, *continuous random variables*, which form the heart of statistics and are used extensively in applied probability as well. It is for such random variables that the calculus prerequisite for this book is needed.

## 6.1 A Random Dart

Imagine that we throw a dart at random at the interval [0,1]. Let $D$ denote the spot we hit. By "at random" we mean that all the points are equally like to be hit. In turn, that means that all subintervals of equal length are equally likely to get hit. For instance, the probability of the dart landing in (0.7,0.8) is the same as for (0.2,0.3), (0.537,0.637) and so on, since they all have length 0.1.

Because of that randomness,

$$P(u \leq D \leq v) = v - u \qquad (6.1)$$

for any case of $0 \leq u < v \leq 1$.

We call $D$ a *continuous* random variable, because its support is a continuum

113

of points, in this case, the entire interval [0,1].

# 6.2   Individual Values Now Have Probability Zero

The first crucial point to note is that

$$P(D = c) = 0 \tag{6.2}$$

for any individual point $c$. This may seem counterintuitive! But it can be seen in a couple of ways:

- Take for example the case $c = 0.3$. Then

$$P(D = 0.3) \leq P(0.29 \leq D \leq 0.31) = 0.02 \tag{6.3}$$

  the last equality coming from (6.1).

  So, $P(D = 0.3) \leq 0.02$. But we can replace 0.29 and 0.31 in (6.3) by 0.299 and 0.301, say, and get $P(D = 0.3) \leq 0.002$. We can continue in this manner, revealing that $P(D = 0.3)$ must be smaller than any positive number, and thus it's actually 0.

- Reason that there are infinitely many points, and if they all had some nonzero probability $w$, say, then the probabilities would sum to infinity instead of to 1; thus they must have probability 0.

Similarly, we will see that (6.2) will hold for any continuous random variable.

It may seem odd at first, but it's very similar to a situation you're quite familiar with from calculus. If I take a slice of some three-dimensional object, say a sphere, and the slice has 0 thickness, then the slice has 0 volume.

In fact, viewed another way, it is not counterintuitive at all. Remember, we have been looking at probability as being the long-run fraction of the time an event occurs, in infinitely many repetitions of our experiment — the "notebook" view. So (6.2) doesn't say that $D = c$ can't occur; it merely says that it happens so rarely that the long-run fraction of occurrence is 0.

Of course, it is still true that continuous random variable models are idealizations, similar to massless, frictionless string in physics. We can only measure the position of the dart to a certain number of decimal places, so it is technically discrete. But modeling it as continuous is a good approximation, and it makes things much easier.

## 6.3 But Now We Have a Problem

But Equation (6.2) presents a problem. In the case of discrete random variables $M$, we defined their distribution via their probability mass function, $p_M$. Recall that Section 5.1 defined this as a list of the values $M$ takes on, together with their probabilities. But that would be impossible in the continuous case — all the probabilities of individual values here are 0.

So our goal will be to develop another kind of function, which is similar to probability mass functions in spirit, but circumvents the problem of individual values having probability 0. To do this, we first must define another key function:

## 6.4 Our Way Out of the Problem: Cumulative Distribution Functions

Here we introduce the notion of a *cumulative distribution function*. It and the concept of a *density* will be used throughout the remainder of this book.

### 6.4.1 CDFs

**Definition 11** *For any random variable $W$ (including discrete ones), its cumulative distribution function (cdf), $F_W$, is defined by*

$$F_W(t) - P(W \le t), -\infty < t < \infty \qquad (6.4)$$

(Please keep in mind the notation. It is customary to use capital F to denote a cdf, with a subscript consisting of the name of the random variable.)

What is $t$ here? It's simply an argument to a function. The function here has domain $(-\infty, \infty)$, and we must thus define that function for every value of $t$. This is a simple point, but a crucial one.

For an example of a cdf, consider our "random dart" example above. We know that, for example for $t = 0.23$,

$$F_D(0.23) = P(D \leq 0.23) = P(0 \leq D \leq 0.23) = 0.23 \qquad (6.5)$$

Also,

$$F_D(-10.23) = P(D \leq -10.23) = 0 \qquad (6.6)$$

and

$$F_D(10.23) = P(D \leq 10.23) = 1 \qquad (6.7)$$

Note that *the fact that D can never be equal to 10.23 or anywhere near it is irrelevant.* $F_D(t)$ is defined for *all* $t$ in $(-\infty, \infty)$, including 10.23! The definition of $F_D(10.23)$ is $P(D \leq 10.23)$, and that probability is 1! Yes, $D$ is *always* less than or equal to 10.23, right?

In general for our dart,

$$F_D(t) = \begin{cases} 0, & \text{if } t \leq 0 \\ t, & \text{if } 0 < t < 1 \\ 1, & \text{if } t \geq 1 \end{cases} \qquad (6.8)$$

The graph of $F_D$ is shown in Figure 6.1.

The cdf of a discrete random variable is defined as in Equation (6.4) too. For example, say $Z$ is the number of heads we get from two tosses of a coin. Then

$$F_Z(t) = \begin{cases} 0, & \text{if } t < 0 \\ 0.25, & \text{if } 0 \leq t < 1 \\ 0.75, & \text{if } 1 \leq t < 2 \\ 1, & \text{if } t \geq 2 \end{cases} \qquad (6.9)$$

For instance,

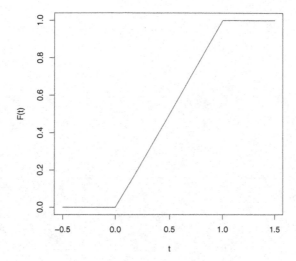

Figure 6.1: Cdf of D

$$
\begin{aligned}
F_Z(1.2) \quad &= \quad P(Z \le 1.2) && (6.10) \\
&= \quad P(Z = 0 \text{ or } Z = 1) && (6.11) \\
&= \quad 0.25 + 0.50 && (6.12) \\
&= \quad 0.75 && (6.13)
\end{aligned}
$$

Note that (6.11) is simply a matter of asking our famous question, "How can it happen?" Here we are asking how it can happen that $Z \le 1.2$. The answer is simple: That can happen if $Z$ is 0 or 1. *The fact that $Z$ cannot equal 1.2 is irrelevant.*

(6.12) uses the fact that $Z$ has a binomial distribution with $n = 2$ and $p = 0.5$. $F_Z$ is graphed Figure 6.2.

The fact that one cannot get a noninteger number of heads is what makes the cdf of $Z$ flat between consecutive integers.

In the graphs you see that $F_D$ in (6.8) is continuous while $F_Z$ in (6.9) has jumps. This is another reason we call random variables such as $D$ *continuous random variables.*

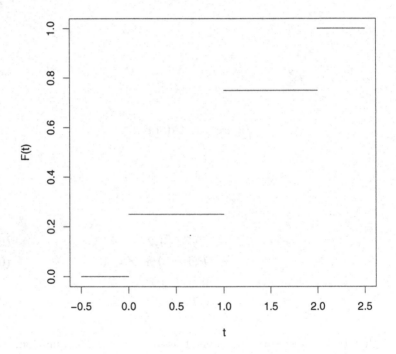

Figure 6.2: Cdf of Z

## 6.4.2  Non-Discrete, Non-Continuous Distributions

Let's modify our dart example above. Say the dart lands at random in [-1,2], and define $D$ as follows. If the dart lands at a point $x$ in [0,1], then set $D$ to $x$. But if the dart lands to the left of 0, define $D$ to be 0, and if it's to the right of 1, set $D$ to 1. Then $D$ is "neither fish nor fowl." It would seem to be continuous, as its support still is a continuum, [0,1]. But it violates (6.2) in the cases $c = 0$ and $c = 1$; for instance, $P(D = 0) = 1/3$, since the subinterval [-1,0] is 1/3 the length of [-1,2] and the dart hits the latter at a random point.

Most of our random variables in this book will be either discrete or continuous, but it's important to know there are other kinds.

## 6.5  Density Functions

Armed with cdfs, let's turn to the original goal, which was to find something for continuous random variables that is similar in spirit to probability mass functions for discrete random variables. That mechanism will be *density functions*.

---

Intuition is key here. Make SURE you develop a good intuitive understanding of density functions, as they are vital in being able to apply probability well. We will use them constantly his book.

---

(The reader may wish to review pmfs in Section 5.1.)

Now think as follows. Look at the 0.25 + 0.50 and 0.75 in (6.12). We see that at jump points $t$, $F_Z(t)$ is a sum of values of $p_Z$ up to that point. This is true in general; for a discrete random variable, its cdf can be calculated by summing its pmf at the set $J_t$ of jump points up through $t$:

$$F_Z(t) = \sum_{j=J} p_Z(j) \qquad (6.14)$$

But recall that in the continuous, i.e., "calculus" world, we integrate instead of sum. (An integral is a limit of sums, which is why the integration symbol $\int$ is shaped like an S.) So, our continuous-case analog of the pmf should be something that integrates to the cdf. That of course is the derivative of the cdf, which is called the *density*:

**Definition 12** *Consider a continuous random variable W. Define*

$$f_W(t) = \frac{d}{dt} F_W(t), \, -\infty < t < \infty \qquad (6.15)$$

*wherever the derivative exists. The function $f_W$ is called the probability density function (pdf), or just the density of W.*

(Again, please keep in mind the notation. It is customary to use lower-case f to denote a density, with a subscript consisting of the name of the random variable.)

But what *is* a density function? First and foremost, it is a tool for finding probabilities involving continuous random variables:

## 6.5.1   Properties of Densities

Equation (6.15) implies

**Property A:**

$$
\begin{aligned}
P(a < W \le b) \;&=\; F_W(b) - F_W(a) &&(6.16) \\
&=\; \int_a^b f_W(t)\, dt &&(6.17)
\end{aligned}
$$

Where does (6.16) come from? Well, $F_W(b)$ is all the probability accumulated from $-\infty$ to b, while $F_W(a)$ is all the probability accumulated from $-\infty$ to a. The difference is then the probability that $X$ is *between* a and b.

(6.17) is just the Fundamental Theorem of Calculus: Integrate the derivative of a function, and you get the original function back again.

Since $P(W = c) = 0$ for any single point c, Property A also means:

**Property B:**

$$P(a < W \le b) = P(a \le W \le b) = P(a \le W < b) = P(a < W < b)$$

$$= \int_a^b f_W(t)\, dt$$

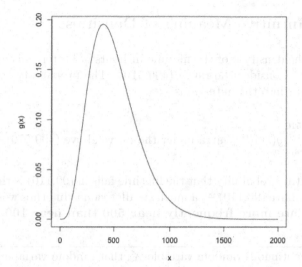

Figure 6.3: Density of battery lifetimes

This in turn implies:

**Property C:**

$$\int_{-\infty}^{\infty} f_W(t)\ dt = 1 \qquad\qquad (6.18)$$

Note that in the above integral, $f_W(t)$ will be 0 in various ranges of t corresponding to values $W$ cannot take on, i.e., outside the support of $W$. For the dart example in Section 6.1, for instance, this will be the case for $t < 0$ and $t > 1$.

Any nonnegative function that integrates to 1 is a density. A density could be increasing, decreasing or mixed. Note too that **a density can have values larger than 1 at some points**, even though it must integrate to 1.

## 6.5.2   Intuitive Meaning of Densities

Suppose the density $g$ of the lifetime in hours of batteries is as depicted in Figure 6.3. Consider the range (490,510). The probability that a battery lifetime falls into this interval is

$$\int_{490}^{510} g(t)\, dt = \text{ area under the curve above (490,510)} \qquad (6.19)$$

Similarly, the probability that the lifetime falls in (90,110) is the area under the curve above (90,110) — a much smaller value. In other words, **battery lifetimes are more frequently near 500 than near 100.** So,

For any continuous random variable $X$, that random variable takes on values more frequently in regions of high density than in low ones.

The reader is probably familiar with histograms. It's the same idea there. Say we have a histogram of exam scores. If the histogram is high around score 68 but low near 86, it means there are a lot more scores near 68 than near 86. Actually, we will see in the next chapter that histograms and densities are closely related.

## 6.5.3   Expected Values

What about $E(W)$? Recall that if $W$ were discrete, we'd have

$$E(W) = \sum_c c\, p_W(c) \qquad (6.20)$$

where the sum ranges over the support of $W$. If for example $W$ is the number of dots we get in rolling two dice, $c$ will range over the values 2,3,...,12.

Again, since in the continuous world we integrate rather than sum, the analog for continuous $W$ is:

**Property D:**

$$E(W) = \int t\, f_W(t)\, dt \qquad (6.21)$$

where here $t$ ranges over the support of $W$, such as the interval $[0,1]$ in the dart case. Again, we can also write this as

$$E(W) = \int_{-\infty}^{\infty} t \, f_W(t) \, dt \tag{6.22}$$

in view of the previous comment that $f_W(t)$ might be 0 for various ranges of $t$.

And of course,

$$E(W^2) = \int_{-\infty}^{\infty} t^2 f_W(t) \, dt \tag{6.23}$$

and in general, similarly to (3.34):

**Property E:**

$$E[g(W)] = \int_t g(t) f_W(t) \, dt \tag{6.24}$$

Most of the properties of expected value and variance stated previously for discrete random variables hold for continuous ones too:

**Property F:**

Equations (3.26), (3.28), (3.36), (4.4) and (4.12) still hold in the continuous case.

## 6.6 A First Example

Consider the density function equal to $2t/15$ on the interval $(1,4)$, 0 elsewhere. Say $X$ has this density. Here are some computations we can do:

$$EX = \int_1^4 t \cdot 2t/15 \, dt = 2.8 \tag{6.25}$$

$$P(X > 2.5) = \int_{2.5}^4 2t/15 \, dt = 0.65 \tag{6.26}$$

$$F_X(s) = \int_1^s 2t/15 \ dt = \frac{s^2-1}{15}, \ s \text{ in } (1,4) \tag{6.27}$$

with $F_X(s$ being 0 for $s < 1$, and 1 for $t > 4$. And

$$\begin{aligned}
Var(X) &= E(X^2) - (EX)^2 \quad \text{(from (4.4)} & (6.28)\\
&= \int_1^4 t^2 2t/15 \ dt - 2.8^2 \quad \text{(from (6.25))} & (6.29)\\
&= 0.66 & (6.30)
\end{aligned}$$

Suppose $L$ is the lifetime of a light bulb (say in years), with the density that $X$ has above. Let's find some quantities in that context:

**Proportion of bulbs with lifetime less than the mean lifetime:**

$$P(L < 2.8) = \int_1^{2.8} 2t/15 \ dt = (2.8^2 - 1)/15 \tag{6.31}$$

(Note that the proportion less than the mean is *not* 0.5 for this distribution.)

**Mean of 1/L:**

$$E(1/L) = \int_1^4 \frac{1}{t} \cdot 2t/15 \ dt = \frac{2}{5} \tag{6.32}$$

**In testing many bulbs, mean number of bulbs that it takes to find two that have lifetimes longer than 2.5:**

Use (5.43) with $r = 2$ and $p = 0.65$.

## 6.7  Famous Parametric Families of Continuous Distributions

Just as with the discrete case, there are a number of parametric families of distributions that have been found useful.

## 6.7.1   The Uniform Distributions

### 6.7.1.1   Density and Properties

In our dart example, we can imagine throwing the dart at the interval (q,r) (so this will be a two-parameter family). Then to be a uniform distribution, i.e., with all the points being "equally likely," the density must be constant in that interval. But it also must integrate to 1 [see (6.18)]. So, that constant must be 1 divided by the length of the interval:

$$f_D(t) = \frac{1}{r - q} \tag{6.33}$$

for $t$ in (q,r), 0 elsewhere.

It easily shown that $E(D) = \frac{q+r}{2}$ and $Var(D) = \frac{1}{12}(r - q)^2$.

The notation for this family is $U(q, r)$.

### 6.7.1.2   R Functions

Relevant functions for a uniformly distributed random variable $X$ on (r,s) are:

- **dunif(x,r,s)**, to find $f_X(x)$

- **punif(q,r,s)**, to find $P(X \leq q)$

- **qunif(q,r,s)**, to find $c$ such that $P(X \leq c) = q$

- **runif(n,r,s)**, to generate $n$ independent values of $X$

As with most such distribution-related functions in R, **x** and **q** can be vectors, so that **punif()** for instance can be used to find the cdf values at multiple points.

By the way, here in the realm of continuous distributions the problem brought up just before (5.36) no longer plagues us. For instance, the reader should make sure to understand why **qunif(0.6,1,4)** is (uniquely) 2.8.

### 6.7.1.3   Example: Modeling of Disk Performance

Uniform distributions are often used to model computer disk requests. A disk consists of a large number of concentric rings, called *tracks*. When a program issues a request to read or write a file, the disk's *read/write head* must be positioned above the track of the first part of the file. This move, which is called a *seek*, can be a significant factor in disk performance in large systems, e.g., a database for a major bank.

If the number of tracks is large, the position of the read/write head, which we'll denote as $X$, is like a continuous random variable, and often this position is modeled by a uniform distribution. This situation may hold most of the time, though after a defragmentation operation, the files tend to be bunched together in the central tracks of the disk, so as to reduce seek time, and $X$ will not have a uniform distribution anymore.

Each track consists of a certain number of *sectors* of a given size, say 512 bytes each. Once the read/write head reaches the proper track, we must wait for the desired sector to rotate around and pass under the read/write head. It should be clear that a uniform distribution is a good model for this *rotational delay*.

For example, suppose in modeling disk performance, we describe the position $X$ of the read/write head as a number between 0 and 1, representing the innermost and outermost tracks, respectively. Say we assume $X$ has a uniform distribution on (0,1), as discussed above. Consider two consecutive positions (i.e., due to two consecutive seeks), $X_1$ and $X_2$, which we'll assume are independent.[1] Let's find $Var(X_1 + X_2)$.

We know from Section 6.7.1.1 that the variance of a $U(0,1)$ distribution is $1/12$. Then by independence,

$$Var(X_1 + X_2) = 1/12 + 1/12 = 1/6 \qquad (6.34)$$

### 6.7.1.4   Example: Modeling of Denial-of-Service Attack

In one facet of computer security, it has been found that a uniform distribution is actually a warning of trouble, a possible indication of a *denial-of-service attack*. Here the attacker tries to monopolize, say, a Web server, by inundating it with service requests. Research has shown [8] that a uniform

---

[1]This assumption may be reasonable if there are a many users of the disk, so that consecutive requests come from different users.

distribution is a good model for IP addresses in this setting.

## 6.7.2 The Normal (Gaussian) Family of Continuous Distributions

These are the famous "bell-shaped curves," so called because their densities have that shape.[2]

### 6.7.2.1 Density and Properties

**Density and Parameters:**

The density for a normal distribution is

$$f_W(t) = \frac{1}{\sqrt{2\pi}\sigma} \, e^{-0.5\left(\frac{t-\mu}{\sigma}\right)^2}, -\infty < t < \infty \qquad (6.35)$$

Again, this is a two-parameter family, indexed by the parameters $\mu$ and $\sigma$, which turn out to be the mean[3] and standard deviation. The notation for it is $N(\mu, \sigma^2)$ (it is customary to state the variance $\sigma^2$ rather than the standard deviation).

### 6.7.2.2 R Functions

Again, R provides functions for the density, cdf, quantile calculation and random number generation, in this case for the normal family:

- dnorm(x, mean = 0, sd = 1)

- pnorm(q, mean = 0, sd = 1)

- qnorm(p, mean = 0, sd = 1)

- rnorm(n, mean = 0, sd = 1)

---

[2] *All that glitters is not gold"* — Shakespeare

Note that other parametric families, notably the Cauchy, also have bell shapes. The difference lies in the rate at which the tails of the distribution go to 0. However, due to the Central Limit Theorem, to be presented in Chapter 9, the normal family is of prime interest.

[3] Remember, this is a synonym for expected value.

Here **mean** and **sd** are of course the mean and standard deviation of the distribution. The other arguments are as in our previous examples.

### 6.7.2.3   Importance in Modeling

The normal family forms the very core of classical probability theory and statistics methodology. Its central role is reflected in the Central Limit Theorem (CLT), which says essentially, that if $X_1, .., , X_n$ are independent and of the same distribution, then the new random variable

$$Y = X_1, .., , X_n \tag{6.36}$$

has an approximately normal distribution.

This family is so important that we have a special chapter on it, Chapter 9, including more on the CLT.

## 6.7.3   The Exponential Family of Distributions

In this section we will introduce another famous parametric family, the family of exponential distributions.[4]

### 6.7.3.1   Density and Properties

The densities in this family have the form

$$f_W(t) = \lambda e^{-\lambda t}, 0 < t < \infty \tag{6.37}$$

This is a one-parameter family of distributions.

After integration, one finds that $E(W) = \frac{1}{\lambda}$ and $Var(W) = \frac{1}{\lambda^2}$.

### 6.7.3.2   R Functions

Relevant functions for a uniformly distributed random variable $X$ with parameter $\lambda$ are

---

[4]This should not be confused, though, with the term *exponential family* that arises in mathematical statistics, which includes exponential distributions but is much broader.

- **dexp(x,lambda)**, to find $f_X(x)$

- **pexp(q,lambda)**, to find $P(X \leq q)$

- **qexp(q,lambda)**, to find $c$ such that $P(X \leq c) = q$

- **rexp(n,lambda)**, to generate $n$ independent values of $X$

### 6.7.3.3 Example: Garage Parking Fees

A certain public parking garage charges parking fees of \$1.50 for the first hour, and \$1 per hour after that. (It is assumed here for simplicity that the time is prorated within each of those defined periods. The reader should consider how the analysis would change if the garage "rounds up" each partial hour.) Suppose parking times $T$ are exponentially distributed with mean 1.5 hours. Let $W$ denote the total fee paid. Let's find $E(W)$ and $Var(W)$.

The key point is that $W$ is a function of $T$:

$$W = \begin{cases} 1.5\, T, & \text{if } T \leq 1 \\ 1.5 + 1 \cdot (T - 1) = T + 0.5, & \text{if } T > 1 \end{cases} \qquad (6.38)$$

That's good, because we know how to find the expected value of a function of a continuous random variable, from (6.24). Defining $g()$ as in (6.38) above, we have

$$EW \;=\; \int_0^\infty g(t)\, \frac{1}{1.5} e^{-\frac{1}{1.5}t}\, dt \qquad (6.39)$$

$$=\; \int_0^1 1.5t\, \frac{1}{1.5} e^{-\frac{1}{1.5}t}\, dt + \int_1^\infty (t + 0.5)\, \frac{1}{1.5} e^{-\frac{1}{1.5}t}\, dt \qquad (6.40)$$

The integration is left to the reader. Or, we can use R's **integrate()** function; see Section 6.9.1.

Now, what about $Var(W)$? As is often the case, it's easier to use (4.4), so we need to find $E(W^2)$. The above integration becomes

$$
\begin{aligned}
E(W^2) &= \int_0^\infty g^2(t)\, f_W(t)dt \\
&= \int_0^1 (1.5t)^2 \, \frac{1}{1.5} e^{-\frac{1}{1.5}t}dt + \int_1^\infty (t+0.5)^2 \, \frac{1}{1.5} e^{-\frac{1}{1.5}t}dt
\end{aligned}
$$

After evaluating this, we subtract $(EW)^2$, giving us the variance of $W$.

### 6.7.3.4  Memoryless Property of Exponential Distributions

One of the reasons the exponential family of distributions is so famous is that it has a property that makes many practical stochastic models mathematically tractable: The exponential distributions are *memoryless*.[5]

What the term *memoryless* means for a random variable $W$ is that for all positive $t$ and $u$

$$
P(W > t + u \mid W > t) = P(W > u) \tag{6.41}
$$

Let's derive this:

$$
\begin{aligned}
P(W > t + u \mid W > t) &= \frac{P(W > t + u \text{ and } W > t)}{P(W > t)} &&(6.42) \\[2mm]
&= \frac{P(W > t + u)}{P(W > t)} &&(6.43) \\[2mm]
&= \frac{\int_{t+u}^\infty \lambda e^{-\lambda s}\, ds}{\int_t^\infty \lambda e^{-\lambda s}\, ds} &&(6.44) \\[2mm]
&= e^{-\lambda u} &&(6.45) \\[2mm]
&= P(W > u) &&(6.46)
\end{aligned}
$$

We say that this means that "time starts over" at time $t$, or that $W$ "doesn't remember" what happened before time $t$.

It is difficult for the beginning modeler to fully appreciate the memoryless property. Let's make it concrete. Say we are driving and reach some

---

[5]The reader may recall that we previously found that geometric distributions are memoryless. It can be shown that family is the only discrete memoryless family, and the exponential distribution family is the only continuous one.

railroad tracks, and we arrive just as a train starts to pass by. One cannot see down the tracks, so we don't know whether the end of the train will come soon or not. The issue at hand is whether to turn off the car's engine. If we leave it on, and the end of the train does not come for a long time, we will be wasting gasoline; if we turn it off, and the end does come soon, we will have to start the engine again, which also wastes gasoline.

Say it's been 2 minutes since the train first started passing, and we would turn off the engine if we knew the train still has 0.5 minutes left. Is that likely, given that the train has already lasted 2 minutes? If the length of the train were exponentially distributed,[6] Equation (6.41) would say that the fact that we have waited 2 minutes so far is of no value at all in predicting whether the train will end within the next 30 seconds. The chance of it lasting at least 30 more seconds right now is no more and no less than the chance it had of lasting at least 30 seconds when it first arrived.

Pretty remarkable!

#### 6.7.3.5 Importance in Modeling

Many distributions in real life have been found to be approximately exponentially distributed. A famous example is interarrival times, such as customers coming into a bank or messages going out onto a computer network. It is used in software reliability studies too.

The exponential family has an interesting (and useful) connection to the Poisson family. This is discussed in Section 6.8.2.

Also, the exponential family is key to the continuous-time version of Markov chains. Here we wait a continuous, random amount of time between jumps between states. Due to the Markov property, this forces the waiting time to be memoryless, thus to have an exponential distribution.

### 6.7.4 The Gamma Family of Distributions

Here is a generalization of the exponential family, also widely used.

---

[6]If there are typically many cars, we can model it as continuous even though it is discrete.

### 6.7.4.1  Density and Properties

Suppose at time 0 we install a light bulb in a lamp, which burns $X_1$ amount of time. We then immediately install a new bulb, which burns for time $X_2$, and so on. Assume the $X_i$ are independent random variables having an exponential distribution with parameter $\lambda$.

Let

$$T_r = X_1 + ... + X_r, \quad r = 1, 2, 3, ... \tag{6.47}$$

Note that the random variable $T_r$ is the time of the $r^{th}$ light bulb replacement. $T_r$ is the sum of $r$ independent exponentially distributed random variables with parameter $\lambda$. The distribution of $T_r$ is called *Erlang*. Its density can be shown to be

$$f_{T_r}(t) = \frac{1}{(r-1)!} \lambda^r t^{r-1} e^{-\lambda t}, \ t > 0 \tag{6.48}$$

This is a two-parameter family.

Again, it's helpful to think in "notebook" terms. Say $r = 8$. Then we watch the lamp for the durations of eight lightbulbs, recording $T_8$, the time at which the eighth burns out. We write that time in the first line of our notebook. Then we watch a new batch of eight bulbs, and write the value of $T_8$ for those bulbs in the second line of our notebook, and so on. Then after recording a very large number of lines in our notebook, we plot a histogram of all the $T_8$ values. The point is then that that histogram will look like (6.48).

We can generalize this by allowing $r$ to take noninteger values, by using a generalization of the factorial function:

$$\Gamma(r) = \int_0^\infty x^{r-1} e^{-x} \, dx \tag{6.49}$$

This is the *gamma function*, well known in classical mathematics. It gives us the gamma family of distributions, more general than the Erlang:

$$f_W(t) = \frac{1}{\Gamma(r)} \lambda^r t^{r-1} e^{-\lambda t}, \ t > 0 \tag{6.50}$$

(Note that $\Gamma(r)$ is merely serving as the constant that makes the density integrate to 1.0. It doesn't have meaning of its own.) This is again a two-parameter family, with $r$ and $\lambda$ as parameters.

A gamma distribution has mean $r/\lambda$ and variance $r/\lambda^2$. In the case of integer $r$, this follows from (6.58) and the fact that an exponentially distributed random variable has mean and variance $1/\lambda$ and variance $1/\lambda^2$. Note again that the gamma reduces to the exponential when $r = 1$.

### 6.7.4.2    Example: Network Buffer

Suppose in a network context (not our ALOHA example), a node does not transmit until it has accumulated five messages in its buffer. Suppose the times between message arrivals are independent and exponentially distributed with mean 100 milliseconds. Let's find the probability that more than 552 ms will pass before a transmission is made, starting with an empty buffer.

Let $X_1$ be the time until the first message arrives, $X_2$ the time from then to the arrival of the second message, and so on. Then the time until we accumulate five messages is $Y = X_1 + ... + X_5$. Then from the definition of the gamma family, we see that Y has a gamma distribution with $r = 5$ and $\lambda = 0.01$. Then

$$P(Y > 552) = \int_{552}^{\infty} \frac{1}{4!} 0.01^5 t^4 e^{-0.01t} \ dt \qquad (6.51)$$

This integral could be evaluated via repeated integration by parts, but let's use R instead:

```
> 1 - pgamma(552,5,0.01)
[1] 0.3544101
```

Note that our parameter $r$ is called *shape* in R, and our $\lambda$ is *rate*.

Again, there are also **dgamma()**, **qgamma()** and **rgamma()**.

### 6.7.4.3    Importance in Modeling

As seen in (6.58), sums of exponentially distributed random variables often arise in applications. Such sums have gamma distributions.

You may ask what the meaning is of a gamma distribution in the case of noninteger $r$. There is no particular meaning, but when we have a real data set, we often wish to summarize it by fitting a parametric family to it, meaning that we try to find a member of the family that approximates our data well.

In this regard, the gamma family provides us with densities which rise near t = 0, then gradually decrease to 0 as t becomes large, so the family is useful if our data seem to look like this. Graphs of some gamma densities are shown in Figure 6.4.

As you might guess from the network performance analysis example in Section 6.7.4.2, the gamma family does arise often in the network context, and in queuing analysis in general. It's also common in reliability analysis.

## 6.7.5   The Beta Family of Distributions

As seen in Figure 6.4, the gamma family is a good choice to consider if our data are nonnegative, with the density having a peak near 0 and then gradually tapering off to the right off to $\infty$. What about data in the range (0,1), or for that matter, any bounded interval?

For instance, say a trucking company transports many things, including furniture. Let $X$ be the proportion of a truckload that consists of furniture. For instance, if 15% of given truckload is furniture, then $X = 0.15$. So here we have a distribution with support in (0,1). The beta family provides a very flexible model for this kind of setting, allowing us to model many different concave-up or concave-down curves on that support.

### 6.7.5.1   Density Etc.

The densities of the family have the following form:

$$\frac{\Gamma(\alpha+\beta)}{\Gamma(\alpha)\Gamma(\beta)}t^{\alpha-1}(1-t)^{\beta-1} \tag{6.52}$$

There are two parameters, $\alpha$ and $\beta$. Figures 6.5 and 6.6 show two possibilities.

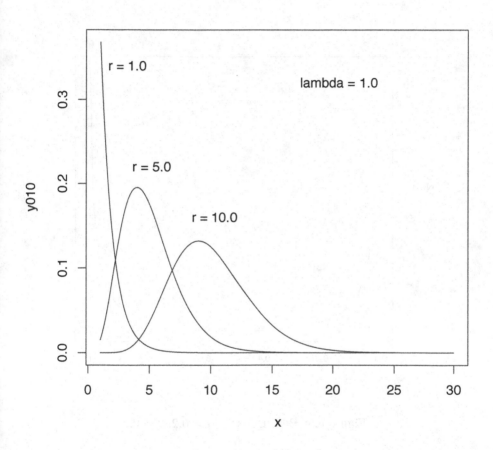

Figure 6.4: Various Gamma Densities

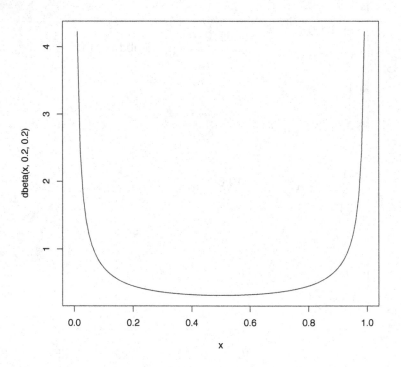

Figure 6.5: Beta Density, $\alpha = 0.2, \beta = 0.2$

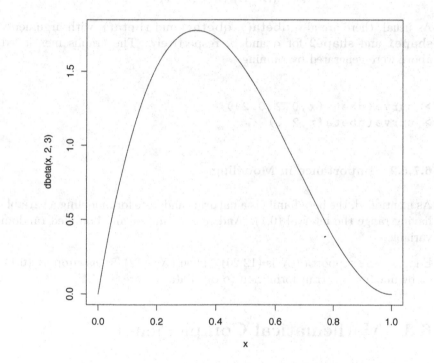

Figure 6.6: Beta Density, $\alpha = 2.0, \beta = 3.0$

The mean and variance can be shown to be

$$\frac{\alpha}{\alpha + \beta} \tag{6.53}$$

and

$$\frac{\alpha\beta}{(\alpha + \beta)^2(\alpha + \beta + 1)} \tag{6.54}$$

As usual, there are also **dbeta()**, **qbeta()** and **rbeta()**, with arguments **shape1** and **shape2** for $\alpha$ and $\beta$, respectively. The graphs mentioned above were generated by running

```
> curve(dbeta(x,0.2,0.2))
> curve(dbeta(x,2,3)
```

### 6.7.5.2   Importance in Modeling

As mentioned, the beta family is a natural candidate for modeling a variable having range the interval (0,1). And it is useful for any bounded random variable.

E.g., say the support of $X$ is (12,20). Then $(X - 12)/8$ has support (0,1), so by making this transformation to our data, we are in (0,1).

## 6.8   Mathematical Complements

### 6.8.1   Hazard Functions

In Section 6.7.3.4, we showed that exponential distributions are memoryless. In the train example there, no matter how long we have waited, the probability of the train ending in the next short period of time is the same. We say that the *hazard function* for exponential distributions is constant. For a nonnegative random variable $X$ with a density, its hazard function is defined to be

$$h_X(t) = \frac{f_X(t)}{1 - F_X(t)}, \ t > 0 \tag{6.55}$$

Intuitively, this is application of conditional probability. Just as

$$P(t < X < t + \delta) \approx f_X(t)\,\delta \qquad (6.56)$$

for small $\delta > 0$,

$$P(t < X < t + \delta \mid X > t) \approx h_X(t)\,\delta \qquad (6.57)$$

For a uniform distribution on (0,1), for instance, the above works out to $1/(1-t)$, an increasing function. The longer we wait, the more likely it is that the event in question will occur soon. This is plausible intuitively, as "time is running out."

It is clear that hazard functions are very useful in reliability applications.

## 6.8.2 Duality of the Exponential Family with the Poisson Family

Suppose the lifetimes of a set of light bulbs are independent and identically distributed *(i.i.d.)*, and consider the following process. At time 0, we install a light bulb, which burns an amount of time $X_1$. Then we install a second light bulb, with lifetime $X_2$. Then a third, with lifetime $X_3$, and so on.

Let

$$T_r = X_1 + ... + X_r \qquad (6.58)$$

denote the time of the $r^{th}$ replacement. Also, let $N(t)$ denote the number of replacements up to and including time $t$. Then it can be shown that if the common distribution of the $X_i$ is exponentially distributed, then the random variables $N(t)$ have a Poisson distribution with mean $\lambda t$. And the converse is true too: If the $X_i$ are independent and identically distributed and $N(t)$ is Poisson for all $t > 0$, then the $X_i$ must have exponential distributions. In summary:

**Theorem 13** *Suppose $X_1, X_2, ...$ are i.i.d. nonnegative continuous random variables. Define*

$$T_r = X_1 + ... + X_r \qquad (6.59)$$

*and*

$$N(t) = \max\{k : T_k \leq t\} \tag{6.60}$$

*Then the distribution of N(t) is Poisson with parameter $\lambda t$ for all t if and only if the $X_i$ have an exponential distribution with parameter $\lambda$.*

In other words, N(t) will have a Poisson distribution if and only if the lifetimes are exponentially distributed.

**Proof**

*"Only if" part:*

The key is to notice that the event $X_1 > t$ is exactly equivalent to $N(t) = 0$. If the first light bulb lasts longer than t, then the count of burnouts at time t is 0, and vice versa. Then

$$
\begin{aligned}
P(X_1 > t) &= P[N(t) = 0] \quad \text{(see above equiv.)} & (6.61) \\
&= \frac{(\lambda t)^0}{0!} \cdot e^{-\lambda t} \quad ((5.46) & (6.62) \\
&= e^{-\lambda t} & (6.63)
\end{aligned}
$$

Then

$$f_{X_1}(t) = \frac{d}{dt}(1 - e^{-\lambda t}) = \lambda e^{-\lambda t} \tag{6.64}$$

That shows that $X_1$ has an exponential distribution, and since the $X_i$ are i.i.d., that implies that all of them have that distribution.

*"If" part:*

We need to show that if the $X_i$ are exponentially distributed with parameter $\lambda$, then for u nonnegative and each positive integer k,

$$P[N(u) = k] = \frac{(\lambda u)^k e^{-\lambda u}}{k!} \tag{6.65}$$

The proof for the case $k = 0$ just reverses (6.61) above. The general case, not shown here, notes first that $N(u) \leq k$ is equivalent to $T_{k+1} > u$. The probability of the latter event can be found by integrating (6.48) from $u$ to $\infty$. One needs to perform $k - 1$ integrations by parts, and eventually one arrives at (6.65), summed from 1 to $k$, as required.

■

The collection of random variables $N(t)$, $t \geq 0$, is called a *Poisson process*. The relation $E[N(t)] = \lambda t$ says that replacements are occurring at an average rate of $\lambda$ per unit time. Thus $\lambda$ is called the *intensity parameter* of the process. It is this "rate" interpretation that makes $\lambda$ a natural indexing parameter in (6.37).

# 6.9 Computational Complements

## 6.9.1 R's integrate() Function

Let's see how to use R to evaluate (6.40).

```
> f <- function(t) exp(-t/1.5) / 1.5
> integrate(function(t) 1.5*t * f(t),0,1)$value +
    integrate(function(t)
        (t+0.5) * f(t),1,Inf)$value
[1] 1.864937
```

As you can see, **integrate()**'s return value is an S3 object, not a number. The latter is available in the **value** component. component.

## 6.9.2 Inverse Method for Sampling from a Density

Suppose we wish to simulate a random variable $X$ with density $f_X$ for which there is no R function. This can be done via $F_X^{-1}(U)$, where $U$ has a $U(0,1)$ distribution. In other words, we call **runif()** and then plug the result into the inverse of the cdf of $X$.

For example, say $X$ has the density $2t$ on (0,1). Then $F_X(t) = t^2$, so $F^{-1}(s) = s^{0.5}$. We can then generate an $X$ as **sqrt(runif(1))**. Here's why:

For brevity, denote $F_X^{-1}$ as G. Our generated random variable is then $Y = G(U)$. Then

$$\begin{aligned} F_Y(t) &= P[G(U) \leq t] & (6.66)\\ &= P[U \leq G^{-1}(t)] & (6.67)\\ &= P[U \leq F_X(t)] & (6.68)\\ &= F_X(t) & (6.69) \end{aligned}$$

(this last coming from the fact that $U$ has a uniform distribution on $(0,1)$).

In other words, $Y$ and $X$ have the same cdf, i.e., the same distribution! This is exactly what we want.

Note that this method, though valid, is not necessarily practical, since computing $F_X^{-1}$ may not be easy.

## 6.9.3   Sampling from a Poisson Distribution

How does **rpois()** work? Section 6.8.2 shows the answer, with $t = 1$. We keep generating exponentially distribution random variables having parameter $\lambda$, until their sum exceeds 1.0. $N(1)$ is then one less than our count of exponential random variables.

And how do we generate those exponential random variables? We simply use the method in Section 6.9.2. $F_X(t) = 1 - \exp(-\lambda t)$, and then solving

$$u = 1 - e^{-\lambda t} \qquad\qquad (6.70)$$

for $t$, we find that $G(s) = -\log(1 - s)/\lambda$.

Thus the method is basically the call (to generate one number)

```
-log(1-runif(1)) / lambda
```

At first, one might think to replace **1-runif(1)** by **runif(1)**, since we get a uniformly distributed random variable either way. However, due to the manner in which uniform random numbers are generated on a computer, **runif()** actually produces numbers in $[0,1)$. The inclusion of 0 there creates a problem when we invoke log.

## 6.10   Exercises

**Mathematical problems:**

**1.** Suppose the random variable $X$ has density $1.5\, t^{0.5}$ on $(0,1)$, $0$ elsewhere. Find $F_X(0.5)$.

**2.** Redo the example in Section 6.7.3.3, under the assumption that the charge for a partial hour is rounded up rather than prorated.

**3.** Consider the network buffer example, Section 6.7.4.2. Find a number $u$ such that there is a 90% chance that the transmission is made before $u$ milliseconds pass.

**4.** Suppose $X_1, ..., X_n$ are i.i.d. U(0,1). Define $R = \max(X_1, ..., X_n)$. Note what this means in notebook terms. For instance, for $n = 3$ and the first two lines,

| notebook line | $X_1$ | $X_2$ | $X_3$ | $R$ |
|---|---|---|---|---|
| 1 | 0.2201 | 0.0070 | 0.5619 | 0.5619 |
| 2 | 0.7688 | 0.2002 | 0.3131 | 0.7688 |

Find the density of $R$. (Hint: First, use the fact that

$$R \leq t \text{ if and only if all } X_i \leq t \qquad (6.71)$$

to find the cdf of $R$.

**5.** Suppose $X_1, ..., X_n$ are independent, with $X_i$ having an exponential distribution with parameter $\lambda_i$. Let $S = \min X_i$. Using an approach similar to that of Problem 4, show that $S$ has an exponential distribution as well, and state the parameter.

**6.** Consider hazard functions, introduced in Section 6.8.1. Show that

$$\int_0^t h_X(s)\ ds = \ln[1 - F_X(t)] \qquad (6.72)$$

Use this to show that the *only* continuous distributions (with densities) that are memoryless are the exponential distributions.

**7.** For the random variable X in Section 6.6, find the skewness (Section 4.5). Feel free to use quantities already computed in the book.

**8.** The *mode* of a density $f_X(t)$ is the value of $t$ for which the density is maximal, i.e., the location of the peak. (Some densities are multimodal, complicating the definition, which we will not pursue here.) Find the mode of a gamma distribution with parameters $r$ and $\lambda$.

**9.** Make precise the following statement, with proper assumptions and meaning: "Most square matrices are invertible." Hint: A matrix is invertible if and only if its determinant is 0. You might consider the $2 \times 2$ case first to guide your intuition.

**Computational and data problems:**

**10.** Consider the network buffer example, Section 6.7.4.2. Suppose interarrival times, instead of having an exponential distribution with mean 100, are uniformly distributed between 60 and 140 milliseconds. Write simulation code to find the new value of (6.51).

**11.** Use R's **integrate()** function to find $E(X^6)$ for an exponential distribution with mean 1.0.

**12.** Say $f_X(t) = 4t^3$ for $0 < t < 1$, 0 elsewhere. Write a function with call form

```
r43(n)
```

to generate **n** random numbers from this distribution. Check it by finding $EX$ analytically and comparing to

```
print(mean(r43(10000)))
```

**13.** Consider a random variable having a beta distribution with both parameters being 0.2. Find the value of its hazard function at 0.75.

**14.** Use R to graph the cdf of the random variable $D$ in Section 6.4.2.

**15.** Follow up on the idea in Section 6.9.3 to write your own version of **rpois()**.

**16.** At the end of Section 6.5.1, it is stated that "a density can have values larger than 1 at some points, even though it must integrate to 1." Give a specific example of this in one of the parametric families in this chapter, showing the family, the parameter value(s) and the point $t$ at which the density is to be evaluated.

**17.** In Section 4.6.1, we introduced the *Markov Inequality*: For a nonnegative random variable $Y$ and positive constant $d$, we have

$$P(Y \geq d) \leq \frac{EY}{d} \qquad (6.73)$$

So we have an upper bound for that probability, but is the bound *tight*? This is a mathetical term meaning, is the bound rather close to the bounded quantity? If not, the bound may not be very useful.

Evaluate the left- and right-hand sides of (6.73) for the case of $Y$ having an exponential distribution with parameter $\lambda$, and use R to graph the discrepancy against $\lambda$, with one curve for each of several values of $d$. (To be useful, the curves should all be on the same graph.)

# Part II

# Fundamentals of Statistics

# Chapter 7

# Statistics: Prologue

*There are three kinds of lies: lies, damned lies and statistics* — variously attributed to Benjamin Disraeli, Mark Twain etc.

Statistics is an application of probability modeling. To get a sampling of what kinds of questions it addresses (pun intended), consider the following problems:

- Suppose you buy a ticket for a raffle, and get ticket number 68. Two of your friends bought tickets too, getting numbers 46 and 79. Let $c$ be the total number of tickets sold. You don't know the value of $c$, but hope it's small, so you have a better chance of winning. How can you estimate the value of $c$, from the data, 68, 46 and 79?

- It's presidential election time. A poll says that 56% of the voters polled support candidate X, with a margin of error of 2%. The poll was based on a sample of 1200 people. How can a sample of 1200 people out of more than 100 million voters have a margin of error that small? And what does the term *margin of error* really mean, anyway?

- A satellite detects a bright spot in a forest. Is it a fire? Or just a reflection of the sun from some shiny object? How can we design the software on the satellite to estimate the probability that this is a fire?

Those who think that statistics is nothing more than adding up columns of numbers and plugging into formulas are badly mistaken. Actually, as

noted, statistics is an application of probability theory. We employ probabilistic models for the behavior of our sample data, and *infer* from the data
accordingly — hence the name, *statistical inference*.

Arguably the most powerful use of statistics is **prediction**, often known
these days as *machine learning*.

This chapter introduces statistics, specifically *sampling* and *point estima-*
*tion*. It is then interwoven with probability modeling in the next few chapters, culminating with the material on prediction in Chapter 15.

# 7.1   Importance of This Chapter

This chapter will be short but of major importance, used throughout the
book from this point on.

A major theme of this book is the use of real datasets, which is rare for
a "math stat" book. The early positioning of this chapter in the book is
in part to give you more practice with expected value and variance before using them in subsequent chapters, but mainly to prepare you for
understanding how our mathematical concepts apply to real data in those
chapters.

# 7.2   Sampling Distributions

We first will set up some infrastructure, which will be used heavily throughout the next few chapters.

## 7.2.1   Random Samples

**Definition 14** *Random variables* $X_1, X_2, X_3, \ldots$ *are said to be i.i.d. if they*
*are independent and identically distributed. The latter term means that* $p_{X_i}$
*or* $f_{X_i}$ *is the same for all* $i$.

Note the following carefully:

> For i.i.d. $X_1, X_2, X_3, \ldots$, we often use X to represent a generic
> random variable having the common distribution of the $X_i$.

**Definition 15** *We say that $X_1, X_2, X_3, ..., X_n$ is a random sample of size $n$ from a population if the $X_i$ are i.i.d. and their common distribution is that of the population.*

(**Please note:** Those numbers $X_1, X_2, X_3, ..., X_n$ collectively form <u>one</u> sample; you should not say anything like "We have $n$ samples.")

A random sample must be drawn in this manner. Say there are $k$ entities in the population, e.g., $k$ people, with values $v_1, ..., v_k$. If we are interested in people's heights, for instance, then $v_1, ..., v_k$ would be the heights of all people in our population. Then a random sample is drawn this way:

  (a) The sampling is done with replacement.

  (b) Each $X_i$ is drawn from $v_1, ..., v_k$, with each $v_j$ having probability $\frac{1}{k}$ of being drawn.

Condition (a) makes the $X_i$ independent, while (b) makes them identically distributed.

If sampling is done without replacement, we call the data a *simple random sample*. Note how this implies lack of independence of the $X_i$. If for instance $X_1 = v_3$, then we know that no other $X_i$ has that value, contradicting independence; if the $X_i$ were independent, knowledge of one should not give us knowledge concerning others.

But we usually assume true random sampling, i.e., with replacement, and will mean that unless explicitly stating otherwise. In most cases, the population is so large, even infinite,[1] that there is no practical distinction, as we are extremely unlikely to sample the same person (or other unit) twice.

Keep this very important point in mind:

> Note most carefully that *each $X_i$ has the same distribution as the population.* If for instance a third of the population, i.e., a third of the $v_j$, are less than 28, then $P(X_i < 28)$ will be 1/3.
>
> If the mean value of X in the population is, say, 51.4, then EX will be 51.4, and so on.
>
> These points are easy to see, but keep them in mind at all times, as they will arise again and again.

---

[1] Infinite? This will be explained shortly.

## 7.3   The Sample Mean — a Random Variable

A large part of this chapter will concern the *sample mean*,

$$\overline{X} = \frac{X_1 + X_2 + X_3 + \ldots + X_n}{n} \tag{7.1}$$

Say we wish to estimate mean household income in a certain state, based on a sample of 500 households. Here $X_i$ is the income of the $i^{th}$ household in our sample; $\overline{X}$ is the mean income in our sample of 500. Note that $\mu$, the mean household income among *all* households in the state, is **unknown**.

A simple yet crucial concept point that $\overline{X}$ is a random variable. Since $X_1, X_2, X_3, \ldots, X_n$ are random variables — we are sampling the population at random — $\overline{X}$ is a random variable too.

### 7.3.1   Toy Population Example

Let's illustrate it with a tiny example. Suppose we have a population of three people, with heights 69, 72 and 70, and we draw a random sample of size 2. As noted, $\overline{X}$ is a random variable. Its support consists of six values:

$$\frac{69 + 69}{2} = 69, \frac{69 + 72}{2} = 70.5, \frac{69 + 70}{2} = 69.5,$$
$$\frac{70 + 70}{2} = 70, \frac{70 + 72}{2} = 71, \frac{72 + 72}{2} = 72 \tag{7.2}$$

So $\overline{X}$ has finite support, only six possible values. It thus is a discrete random variable, and its pmf is given by 1/9, 2/9, 2/9, 1/9, 2/9 and 1/9, respectively. So,

$$p_{\overline{X}}(69) = \frac{1}{9}, \quad p_{\overline{X}}(70.5) = \frac{2}{9}, \quad p_{\overline{X}}(69.5) = \frac{2}{9},$$
$$p_{\overline{X}}(70) = \frac{1}{9}, \quad p_{\overline{X}}(71) = \frac{2}{9}, \quad p_{\overline{X}}(72) = \frac{1}{9} \tag{7.3}$$

Viewing it in "notebook" terms, we might have, in the first three lines:

| notebook line | $X_1$ | $X_2$ | $\overline{X}$ |
|---|---|---|---|
| 1 | 70 | 70 | 70 |
| 2 | 69 | 70 | 69.5 |
| 3 | 72 | 70 | 71 |

Again, the point is that all of $X_1$, $X_2$ and $\overline{X}$ are random variables.

## 7.3.2   Expected Value and Variance of $\overline{X}$

Now, returning to the case of general $n$ and our sample $X_1, ..., X_n$, since $\overline{X}$ is a random variable, we can ask about its expected value and variance. Note that in notebook terms, these are the long-run mean and variance of the values in the $\overline{X}$ column above.

Let $\mu$ denote the population mean. Remember, each $X_i$ is distributed as is the population, so $EX_i = \mu$. Again in notebook terms, this says that the long-run average in the $X_1$ column will be $\mu$. (The same will be true for the $X_2$ column and so on.)

This then implies that the expected value of $\overline{X}$ is also $\mu$. Here's why:

$$
\begin{aligned}
E(\overline{X}) &= E\left[\frac{1}{n}\sum_{i=1}^{n} X_i\right] && (\text{def. of } \overline{X}) \\
&= \frac{1}{n}E\left(\sum_{i=1}^{n} X_i\right) && (\text{for const. c, } E(cU) = cEU) \\
&= \frac{1}{n}\sum_{i=1}^{n} EX_i && (E[U+V] = EU + EV) \\
&= \frac{1}{n}n\mu && (EX_i = \mu) \\
&= \mu && (7.4)
\end{aligned}
$$

Moreover, the variance of $\overline{X}$ is $1/n$ times the population variance:

$$
\begin{aligned}
Var(\overline{X}) &= Var\left[\frac{1}{n}\sum_{i=1}^{n}X_i\right] \\
&= \frac{1}{n^2}Var\left(\sum_{i=1}^{n}X_i\right) \\
&= \frac{1}{n^2}\sum_{i=1}^{n}Var(X_i) \\
&= \frac{1}{n^2}n\sigma^2 \\
&= \frac{1}{n}\sigma^2 \qquad\qquad (7.5)
\end{aligned}
$$

(The second equality comes from the relation $Var(cU) = c^2 Var(U)$, while the third comes from the additivity of variance for independent random variables.)

*This derivation plays a crucial role in statistics*, and you in turn can see that the independence of the $X_i$ played a crucial role in the derivation. This is why we assume sampling with replacement.

## 7.3.3   Toy Population Example Again

Let's verify (7.4) and (7.5) for toy population in Section 7.3.1. The population mean is

$$
\mu = (69 + 70 + 72)/3 = 211/3 \qquad\qquad (7.6)
$$

Using (3.19) and (7.3), we have

$$
E\overline{X} = \sum_{c}c\, p_{\overline{X}}(c) = 69\cdot\frac{1}{9} + 69.5\cdot\frac{2}{9} + 70\cdot\frac{1}{9} + 70.5\cdot\frac{2}{9} + 71\cdot\frac{2}{9} + 72\cdot\frac{1}{9} = 211/3 \qquad (7.7)
$$

So, (7.4) is confirmed. What about (7.5)?

First, the population variance is

$$
\sigma^2 = \frac{1}{3}\cdot(69^2 + 70^2 + 72^2) - (\frac{211}{3})^2 = \frac{14}{9} \qquad\qquad (7.8)
$$

The variance of $\overline{X}$ is

$$Var(\overline{X}) \;=\; E(\overline{X}^2) - (E\overline{X})^2 \tag{7.9}$$

$$=\; E(\overline{X}^2) - (\frac{211}{3})^2 \tag{7.10}$$

Using (3.34) and (7.3), we have

$$E(\overline{X}^2) = \sum_c c^2\, p_{\overline{X}}(c) = 69^2 \cdot \frac{1}{9} + 69.5^2 \cdot \frac{2}{9} + 70^2 \cdot \frac{1}{9} + 70.5^2 \cdot \frac{2}{9} + 71^2 \cdot \frac{2}{9} + 72^2 \cdot \frac{1}{9} \tag{7.11}$$

The reader should now wrap things up and confirm that (7.9) does work out to $(14/9)\,/\,2 = 7/9$, as claimed by (7.5) and (7.8).

## 7.3.4 Interpretation

Now, let's step back and consider the significance of the above findings (7.4) and (7.5):

(a) Equation (7.4) tells us that although some samples give us an $\overline{X}$ that is too high, i.e., that overestimates $\mu$, while other samples give us an $\overline{X}$ that is too low, on average $\overline{X}$ is "just right."

(b) Equation (7.5) tells us that for large samples, i.e., large $n$, $\overline{X}$ doesn't vary much from sample to sample.

If you put (a) and (b) together, it says that for large n, $\overline{X}$ is probably pretty accurate, i.e., pretty close to the population mean $\mu$. So, the story of statistics often boils down to asking, "Is the variance of our estimator small enough?" You'll see this in the coming chapters, but will give a preview later below.

## 7.3.5 Notebook View

Let's take a look at all this using our usual notebook view. But instead of drawing the notebook, let's generate it by simulation. Say the distribution of $X$ in the population is Poisson with mean 2.5, and that $n = 15$.

We will generate that notebook as a matrix, with one row of the matrix stored the $n$ values of $X$ for one sample, i.e., one row of the notebook. Let's perform the experiment — again, which consists of drawing a sample of $n$ values of $X$ — 5000 times, thus 5000 lines in the notebook.

Here is the code:

```
gennb <- function()
{
    xs <- rpois(5000*15,2.5)   # the Xs
    nb <- matrix(xs,nrow=5000)  # the notebook
    xbar <- apply(nb,1,mean)
    cbind(nb,xbar)   # add Xbar col to notebook
}
```

The **apply()** function is described in more detail in Section 7.12.1.1. In the case here, the call instructs R to call the **mean()** function on each row of **nb**. That amounts to saying we compute $\overline{X}$ for each line in the notebook. So, **xbar** will consist of the 5000 row means, i.e., the 5000 values of $\overline{X}$.

Recall that the variance of a Poisson random variable is the same as the mean, in this case 2.5. Then (7.4) and (7.5) imply that the mean and variance of the $\overline{X}$ column of the notebook should be approximately 2.5 and $2.5/15 = 0.1667$. (Only approximately, since we are looking at only 5000 rows of the notebook, not infinitely many.) Let's check it:

```
> mean(nb[,16])
[1] 2.499853
> var(nb[,16])
[1] 0.1649165
```

Sure enough!

## 7.4   Simple Random Sample Case

What if we sample without replacement? The reader should make sure to understand that (7.4) still holds completely. The additivity of $E()$ holds even if the summands are not independent. And the distribution of the $X_i$ is still the population distribution as in the with-replacement case. (The reader may recall the similar issue in Section 4.4.3.)

What does change is the derivation (7.5). The summands are no longer independent, so variance is no longer additive. That means that covariance terms must be brought in, as in (4.31), and though one might proceed as before in a somewhat messier set of equations, for general statistical procedures this is not possible. So, the independence assumption is ubiquitous.

Actually, simple random sampling does yield smaller variances for $\overline{X}$. This is good, and makes intuitive sense — we potentially sample a greater number of different people. So in our toy example above, the variance will be smaller than the 14/9 value obtained there. The reader should verify this.

As mentioned, in practice the with/without-replacement issue is moot. Unless the population is tiny, the chances of sampling the same person twice is minuscule.

## 7.5 The Sample Variance

As noted, we use the sample mean $\overline{X}$ to estimate the population mean $\mu$. $\overline{X}$ is a function of the $X_i$. What other function of the $X_i$ can we use to estimate the population *variance* $\sigma^2$?

Let X denote a generic random variable having the distribution of the $X_i$, which, note again, is the distribution of the population. Because of that property, we have

$$Var(X) = \sigma^2 \quad (\sigma^2 \text{ is the population variance}) \qquad (7.12)$$

Recall that by definition

$$Var(X) = E[(X - EX)^2] \qquad (7.13)$$

### 7.5.1 Intuitive Estimation of $\sigma^2$

Let's estimate $Var(X) = \sigma^2$ by taking sample analogs in (7.13). The correspondences are shown in Table 7.1.

The sample analog of $\mu$ is $\overline{X}$. What about the sample analog of the "$E()$"? Well, since $E()$ averages over the whole population of $X$s, the sample analog is to average over the sample. So, our sample analog of (7.13) is

$$s^2 = \frac{1}{n} \sum_{i=1}^{n} (X_i - \overline{X})^2 \qquad (7.14)$$

Table 7.1: Population and Sample Analogs

| pop. entity | samp. entity |
|---|---|
| EX | $\overline{X}$ |
| X | $X_i$ |
| E[] | $\frac{1}{n}\sum_{i=1}^{n}$ |

In other words, just as it is natural to estimate the population mean of $X$ by its sample mean, the same holds for $Var(X)$:

> The population variance of $X$ is the mean squared distance from $X$ to its population mean, as $X$ ranges over all of the population. Therefore it is natural to estimate $Var(X)$ by the average squared distance of $X$ to its sample mean, among our sample values $X_i$, shown in (7.14).

We use $s^2$ as our symbol for this estimate of population variance.[2]

### 7.5.2    Easier Computation

By the way, it can be shown that (7.14) is equal to

$$\frac{1}{n}\sum_{i=1}^{n}X_i^2 - \overline{X}^2 \qquad (7.15)$$

This is a handy way to calculate $s^2$, though it is subject to more roundoff error. Note that (7.15) is a sample analog of (4.4).

### 7.5.3    Special Case: X Is an Indicator Variable

We often have data in the form of an indicator variable. For instance, $X$ may be 1 if the person is a student, 0 otherwise. Let's look at $\overline{X}$ and $s^2$ in such a setting.

---

[2]Though I try to stick to the convention of using only capital letters to denote random variables, it is conventional to use lower case in this instance.

From Section 4.4, we know that $\mu = p$ and $\sigma^2 = p(1-p)$, where $p = P(X = 1)$. As usual, keep in mind that these are population quantities.

The natural estimator of $p$ is $\widehat{p}$, the sample proportion of 1s.[3] Note that is actually is $\overline{X}$! This follows from the fact that the numerator of the latter is a sum of 0s and 1s, thus just the count of 1s.

That also has implications for (7.5.2). Since $X_i^2 = X_i$, then (7.5.2) is

$$s^2 = \overline{X} - \overline{X}^2 = \overline{X}(1 - \overline{X}) = \widehat{p}(1 - \widehat{p}) \qquad (7.16)$$

We'll see later that this is very relevant for the election survey at the beginning of this chapter.

## 7.6  To Divide by n or n-1?

It should be noted that it is common to divide by $n - 1$ instead of by $n$ in (7.14). In fact, almost all textbooks divide by $n - 1$ instead of $n$. Clearly, unless $n$ is very small, the difference will be minuscule; such a small difference is not going to affect any analyst's decisionmaking. But there are a couple of important conceptual questions here:

- Why do most people (and R, in its **var()** function) divide by $n - 1$?

- Why do I choose to use $n$?

The answer to the first question is that (7.14) is what is called **biased downwards**. What does this mean?

### 7.6.1  Statistical Bias

**Definition 16** *Suppose we wish to estimate some population quantity $\theta$, using an estimator $\widehat{\theta}$ computed from our sample data. $\widehat{\theta}$ is called unbiased if*

$$E\widehat{\theta} = \theta \qquad (7.17)$$

---

[3]The "hat" symbol,⌢ is a standard way to name an estimator.

Otherwise it is *biased*. The amount of bias is

$$E\widehat{\theta} - \theta \tag{7.18}$$

(7.4) shows that $\overline{X}$ is an unbiased estimate of $\mu$. However, it can be shown (Exercise 5) that

$$E(s^2) = \frac{n-1}{n}\sigma^2 \tag{7.19}$$

In notebook terms, if we were to take many, many samples, one per line in the notebook, in the long run the average of all of our $s^2$ values would be slightly smaller than $\sigma^2$.

This bothered the early pioneers of statistics, so as a "fudge factor" they decided to divide by $n-1$ to make the sample variance an unbiased estimator of $\sigma^2$. Their definition of $s^2$ is

$$s^2 = \frac{1}{n-1}\sum_{i=1}^{n}(X_i - \overline{X})^2 \tag{7.20}$$

This bias avoidance is why W. Gossett defined his now-famous Student-t distribution using (7.20), with a divisor of $n-1$ instead of $n$. One additional nice aspect of this approach is that if one uses the $n-1$ divisor, $s^2$ has a chi-squared distribution if the population distribution is normal. But he could have just as easily defined it as (7.14). There is nothing inherently wrong with small, nonzero bias.

Moreover, even though $s^2$ is unbiased under Gossett's definition, his $s$ itself is still biased downward (Exercise 6). And since $s$ itself is what we (this book and all others) use in forming *confidence intervals* (Chapter 10), one can see that insisting on unbiasedness is a losing game.

I choose to use (7.14), dividing by $n$, because of Table 7.1; it's very important that students understand this idea of sample analogs. Another virtue of this approach is that it is in a certain sense more consistent; when dealing with binary data (Section 10.6), it is standard statistical practice in all books to divide by $n$ rather than $n-1$.

The idea of a confidence interval is central to statistical inference. But actually, you already know about it — from the term *margin of error* in news reports about opinion polls.

# 7.7 The Concept of a "Standard Error"

As noted, $\overline{X}$ is a random variable; its value is different for each sample, and since the sample is random, that makes $\overline{X}$ a random variable.

From (7.5), the standard deviation of that random variable is $\sigma/\sqrt{n}$. Since $\overline{X}$ is a statistical estimator, we say call its standard deviation $\sigma/\sqrt{n}$ the *standard error of the estimate*, or simply the *standard error*.

Also, we of course usually don't know the population standard deviation $\sigma$, so we use $s$ instead. In other words, the standard error of $\overline{X}$ is actually defined to be $s/\sqrt{n}$. In later chapters, we'll meet other estimators of other things, but the standard error will still be defined to be the estimated standard deviation of the estimator, with significance as seen below.

Note that due to (7.16), in the indicator variable case, the standard error of $\widehat{p}$ is

$$\sqrt{\widehat{p}(1-\widehat{p})/n} \tag{7.21}$$

Why give this quantity a special name (which will come up repeatedly in later chapters and in R output)? The answer is that it is key to giving us an idea as to whether an estimate is likely close to the true population value. For that reason, news reports about election polls report the *margin of error*, which you'll see later is usually about double the standard error. We'll go into that in depth with more powerful tools in Chapter 10, but we can gain some insight by applying Chebychev's Inequality, (4.21):

Recall our comment on that equation at the time, taking $c = 3$ in the inequality:

$X$ strays more than, say, 3 standard deviations from its mean at most only 1/9 of the time.

Applying this to the random variable $\overline{X}$, we have

In at least 8/9 of all possible samples, $\overline{X}$ is within 3 standard errors, i.e., $3s/\sqrt{n}$, of the population mean $\mu$.

Interesting, but it will turn out that this assessment of accuracy of $\overline{X}$ is rather crude. We'll return to this matter in Chapter 9.

## 7.8   Example: Pima Diabetes Study

Consider the famous Pima diabetes study [15].

```
names(pima) <- c('NPreg','Gluc','BP','Thick',
   'Insul','BMI','Genet','Age','Diab')
```

The data consist of 9 measures on 767 women. Here are the first few reoords:

```
> pima <-
    read.csv('pima-indians-diabetes.data',header=T)
> head(pima)
  NPreg Gluc BP Thick Insul  BMI Genet Age Diab
1     6  148 72    35     0 33.6 0.627  50    1
2     1   85 66    29     0 26.6 0.351  31    0
3     8  183 64     0     0 23.3 0.672  32    1
4     1   89 66    23    94 28.1 0.167  21    0
5     0  137 40    35   168 43.1 2.288  33    1
6     5  116 74     0     0 25.6 0.201  30    0
```

Here **Diab** uses 1 to code diabetic, 0 for nondiabetic. Let $\mu_1$ and $\sigma_1^2$ denote the population mean and variance of Body Mass Index **BMI** among diabetics, with $\mu_0$ and $\sigma_0^2$ representing the nondiabetics. Let's find their sample estimates:

```
> tapply(pima$BMI,pima$Diab,mean)
       0        1
30.30420 35.14254
> tapply(pima$BMI,pima$Diab,var)
       0        1
59.13387 52.75069
```

There will be details on how **tapply()** works in Section 7.12.1.3. In short, in the first call above, we are instructing R to partition the **BMI** vector according to the **Diab** vector, i.e., create two subvectors of **BMI**, for the diabetics and nondiabetics. We then apply **mean()** to each subvector.

Let's find "n" as well:

```
> tapply(pima$BMI,pima$Diab,length)
  0   1
500 268
```

Call these $n_1$ and $n_0$.

The diabetics do seem to have a higher BMI. We must keep in mind, though, that some or all of the difference may be due to sampling variation. Again, we'll treat that issue more formally in Chapter 10, but let's look a bit at it now.

Here are the standard errors of the two sample means:

```
> sqrt(52.75069/268)
[1] 0.4436563
> sqrt(59.13387/500)
[1] 0.3439008
```

So for the diabetics, for instance, our estimate 35.14254 is likely within about $3 \times 0.4436563 = 1.330969$ of the true population mean BMI $\mu$.[4]

Let's also look at estimating the difference $\mu_1 - \mu_0$. The natural estimate would be $\overline{U} - \overline{V}$, where those two quantities are the diabetic and nondiabetic sample means. A key point is that these two sample means, being from separate data, are independent. Thus, using (4.35), we have that $W = \overline{U} - \overline{V}$ has variance $\sigma_1^2/n_1 + \sigma_0^2/n_0$

In other words:

Standord Error of the Difference of Two Sample Means

The standard error of $\overline{U} - \overline{V}$ is

$$\sqrt{\frac{s_1^2}{n_1} + \frac{s_2^2}{n_2}} \qquad (7.22)$$

Putting this all together, we see that $W$, our estimator of $\mu_1 - \mu_0$, has standard error

$$\sqrt{52.75069/268 + 59.13387/500} \qquad (7.23)$$

or about 0.56. Since our estimate of the difference between the two population means is about 4.8, our informal "3 standard errors" guideline would seem to say there is a substantial difference between the two populations.

---

[4]There are issues with the use of the word *likely* here, to be discussed in Chapter 10.

## 7.9    Don't Forget: Sample $\neq$ Population!

It is quite clear that the sample mean is not the same as the population mean. On the contrary, we use the former to estimate the latter. But in my experience, in complex settings, this basic fact is often overlooked by learners of probability and statistics. A simple but important point, to be kept at the forefront of one's mind.

## 7.10    Simulation Issues

Monte Carlo simulation is fundamentally a sampling operation. If we replicate our experiment **nreps** times, the latter is our sample size.

### 7.10.1    Sample Estimates

Look at the code in Section 3.9, for instance. Our sample size is 10000; $X_i$ is the value of **passengers** at the end of iteration $i$ of the **for** loop; and $\overline{X}$ is the value of **total/nreps**.

Moreover, to get the corresponding standard error, we would compute $s$ from all the **passengers** values, and divide by the square root of **nreps**.

This then would be a partial answer to the question, "How long should we run the simulation?", raised in Section 2.7. The question is then whether the standard error is small enough. This will be addressed further in Chapter 10.

### 7.10.2    Infinite Populations?

Also, what about our mysterious comment in Section 7.2.1 above that the sampled population might be infinite? Here's why:

As noted in Section 2.3, the R function **runif()** can be used as the basis for generating random numbers for a simulation. In fact, even **sample()** calls something equivalent to **runif()** internally. But, in principle, the function **runif()** draws from infinitely many values, namely all the values in the continuum (0,1). Hence our infinite population.

In actuality, that is not quite true. Due to the finite precision of computers, **runif()** actually can have only finitely many values. But as an approximation, we think of **runif()** truly taking on all the values in the continuum. This is true for all continuous random variables, which we pointed out before are idealizations.

## 7.11 Observational Studies

The above formulation of sampling is also rather idealized. It assumes a well-defined population, from which each unit is equally likely to be sampled. In real life, though, things are often not so clearcut.

In Chapter 15, for instance, we analyze data on major league baseball players, and apply statistical inference procedures based on the material in the current chapter. The player data is for a particular year, and our population is the set of all major league players, past, present and future. But here, no physical sampling occurred; we are implicitly assuming that our data *act like* a random sample from that population.

That in turn means that there was nothing special about our particular year. A player in our year, for instance, is just as likely to weigh more than 220 pounds than in previous years. This is probably a safe assumption, but at the same time it probably means we should restrict our (conceptual) population to recent years; back in the 1920s, players probably were not as big as those we see today.

The term usually used for settings like that of the baseball player data is *observational studies*. We passively *observe* our data rather than obtaining it by physically sampling from the population. The careful analyst must consider whether his/her data are representative of the conceptual population, versus subject to some bias.

## 7.12 Computational Complements

### 7.12.1 The *apply() Functions

R's functions **apply()**, **lapply()**, **tapply()** and so on are workhorses in R. They should be mastered in order to do effective data science, and thus it is worth spending a bit of time on the matter here. Fortunately, they are simple and easy to learn.

### 7.12.1.1   R's apply() Function

The general form is[5]

```
apply(m,d,f)
```

where **m** is a matrix or data frame, **d** is a direction (1 for rows, 2 for columns) and **f** is a function to be called on each row or column. (If **m** is a data frame, the case **d = 2** is not allowed, and one must use **lapply()**, explained in the next section.)

On page 156, we had **nb, 1** and R's built-in **mean()** serving as **m, d** and **f**, respectively. More often, one uses a function one writes oneself.

### 7.12.1.2   The lapply() and sapply() Function

Here the 'l' stands for "list"; we apply the same function to all elements of an R list. For example:

```
> l <- list(u='abc',v='de',w='f')
> lapply(l,nchar)
$u
[1] 3

$v
[1] 2

$w
[1] 1
```

R's built-in function **nchar()** returns the length of a character string. As you can see, R returned another R list with the answers. We can simplify by calling **unlist()** on the result, returning the vector (3,2,1). Or if we know it can be converted to a vector, we can use **sapply()**:

```
> sapply(l,nchar)
u v w
3 2 1
```

An R data frame, though matrix-like in various senses, is actually implemented as an R list, one element per column. So, though as mentioned

---

[5]This is not quite the most general form, but sufficient for our purposes here.

above **apply**() cannot be used with **d=2** for data frames, we can use **lapply**() or **sapply**().

For instance:

```
> x <- data.frame(a=c(1,5,0,1),b=c(6,0,0,1))
> x
  a b
1 1 6
2 5 0
3 0 0
4 1 1
> count1s <- function(z) sum(z == 1)
> sapply(x,count1s)
a b
2 1
```

### 7.12.1.3  The split() and tapply() Functions

Another function in this family is **tapply**(), along with the related **split**(). The latter splits a vector into groups according to levels of an R factor. The former does this too, but applies a user-designated function, such as **mean**(), to each group.

Consider this code from Section 7.8:

```
tapply(pima$BMI,pima$Diab,mean)
tapply(pima$BMI,pima$Diab,var)
tapply(pima$BMI,pima$Diab,length)
```

In the first call, we asked R to find the mean of **BMI** in each of the subgroups defined by **Diab**. If we had merely called **split**(),

```
split(pima$BMI,pima$Diab)
```

we would have gotten two subvectors, corresponding to diabetic and non-diabetic, forming a two-element R list.

In the second call, we did the same for sample variance. (Note: The R function **var**() uses the standard $n-1$ divisor, but as noted, the difference is minuscule.) Finally, we found the size of each group by applying **length**() to the group subvectors.

Keep these functions in mind; they will come in handy!

## 7.12.2   Outliers/Errors in the Data

Let's take a look at the glucose column of the Pima data (Section 7.8).

```
> hist(pima$Gluc)
```

The histogram is shown in Figure 7.1. It seems some women in the study had glucose at or near 0. Let's look more closely:

```
> table(pima$Gluc)
```

```
   0   44   56   57   61   62   65   67   68   71   72   73   74
  75   76
   5    1    1    2    1    1    1    1    3    4    1    3    4
   2    2
  77   78   79   80   81   82   83   84   85   86   87   88   89
  90   91
   2    4    3    6    6    3    6   10    7    3    7    9    6
  11    9
...
```

So there were five cases of 0 glucose, physiologically impossible. Let's check for other invalid cases:

```
> apply(pima,2,function(x) sum(x == 0))
NPreg  Gluc    BP Thick Insul   BMI Genet    Age
  111     5    35   227   374    11     0      0
 Diab
  500
```

What happened here? We asked R to go column-by-column in **pima**, calculating the number of 0 values in the column. We accomplished the latter by defining a function, formal argument **x**, that first executes **x** == 0, yielding a vector of **TRUE** and **FALSE** values. R, as is common among programming languages, treats those as 1s and 0s. By then applying **sum()**, we get the count of 0s.

In R, the code for missing values is 'NA'. Various R functions will skip over NA values or otherwise take special action. So, we should recode (not including the first column, number of pregnancies, and the last column, an indicator for diabetes):

```
> pima.save <- pima   # will use again below
> pm19 <- pima[,-c(1,9)]
> pm19[pm19 == 0] <- NA
```

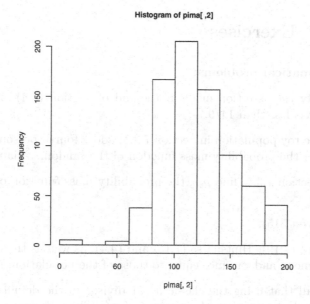

Figure 7.1: Glucose

```
> pima[,-c(1,9)] <- pm19
```

Let's see how much things change without the 0 values:

```
> mean(pima$Gluc)
[1] 120.8945
> mean(pima.save$Gluc)
[1] NA
> mean(pima.save$Gluc,na.rm=TRUE)
[1] 121.6868
```

Some R functions automatically skip over NAs, while others do so only if we request it. Here we did the latter by setting "NA remove" to **TRUE**.

We see that the NAs made only a small difference here, but in some settings they can be substantial.

# 7.13   Exercises

**Mathematical problems:**

**1.** Verify the assertion made at the end of Section (7.4), stating that a variance is less than $14/9$.

**2.** In the toy population in Section 7.3.1, add a fourth person of height 65. Find $p_{\overline{X}}$, the probability mass function of the random variable $\overline{X}$.

**3.** In Section 7.3.1, find $p_{s^2}$, the probability mass function of the random variable $s^2$.

**4.** Derive (7.15).

**5.** Derive (7.19). Hints: Use (7.15) and (4.4), as well as the fact that each $X_i$ has mean and variance equal to those of the population, i.e., $\mu$ and $\sigma^2$.

**6.** Recall that using the classic $n - 1$ divisor in the definition of $s^2$, the latter is then an unbiased estimator of $\sigma^2$. Show, however, that implies that $s$ is then biased; specifically, $Es < \sigma$ (unless $\sigma = 0$). (Hint: Make use of the unbiasedness of $s^2$ and (4.4).)

**Computational and data problems:**

**7.** In Section 7.12.2, we found a slight change in the sample mean after removing the invalid 0s. Calculate how much the sample variance changes.

**8.** Make the code in Section 7.12.2 that replaced 0 values by NAs general, by writing a function with call form

```
zerosToNAs(m,colNums)
```

Here **m** is the matrix or data frame to be converted, and **colNums** is the vector of column numbers to be affected. The return value will be the new version of **m**.

# Chapter 8

# Fitting Continuous Models

*All models are wrong, but some are useful.* — George Box (1919-2013), pioneering statistician

One often models one's data using a parametric family, as in Chapters 5 and 6. This chapter introduces this approach, involving core ideas of statistics, closely related to each other:

- *Why* might we want to fit a parametric model to our sample data?

- *How* do we fit such a model, i.e., how do we estimate the population parameters from our sample data?

- And what constitutes a good fit?

Our focus here will be on fitting parametric density models, thus on continous random variables. However, the main methods introduced, the Method of Moments and Maximum Likelihood Estimation, do apply to discrete random variables as well.

## 8.1   Why Fit a Parametric Model?

Denote our data by $X_1, ..., X_n$. It is often useful to fit a parametric density model to the data. One might ask, though, why bother with a model?

Isn't, say a histogram (see below) enough to describe the data? There are a couple of answers to this:

- In our first example below, we will fit the gamma distribution. The gamma is a two-parameter family, and it's a lot easier to summarize the data with just two numbers, rather than the 20 bin heights in the histogram.

- In many applications, we are working with large systems consisting of dozens of variables. In order to limit the complexity of our model, it is desirable to have simple models of each component.

  For example, in models of *queuing systems* [3]. If things like service times and job interarrival times can be well modeled by an exponential distribution, the analysis may simplify tremendously, and quantities such as mean job waiting times can be easily derived.[1]

## 8.2 Model-Free Estimation of a Density from Sample Data

Before we start with parametric models, let's see how can estimate a density function without them. In addition to providing a contrast to the parametric models, this will introduce central issues that will arise again in regression models and machine learning, Chapter 15.

How can we estimate a population density from our sample data? It turns out that the common histogram, so familiar from chemistry class instructors' summaries of the "distribution" of exam scores, is actually a density estimator! That goes back to our point in Section 6.5.2, which in paraphrased form is:

> Although densities themselves are not probabilities, they do tell us which regions will occur often or rarely.

That is exactly what a histogram tells us.

### 8.2.1   A Closer Look

Let $X_1$, $X_2$, ..., $X_n$ denote our data, a random sample from a population. Say bin $i$ in a histogram covers the interval $(c, c + w)$. Let $N_i$ denote the

---

[1]Even nonexponential times can be handled, e.g., through the *method of stages*.

number of data points falling into the bin. This quantity has a binomial distribution with $n$ trials and success probability

$$p = P(c < X < c + w) = \text{ area under } f_X \text{ from to to c+w} \tag{8.1}$$

where $X$ has the distribution of our variable in the population. If $w$ is small, then this implies

$$p \approx w \, f_X(c) \tag{8.2}$$

But since $p$ is the probability of an observation falling into this bin, we can estimate it by

$$\widehat{p} = \frac{N_i}{n} \tag{8.3}$$

So, we have an estimate of $f_X$!

$$\widehat{f}_X(c) = \frac{N_i}{wn} \tag{8.4}$$

So, other than a constant factor $q/(wn)$, our histogram, which plots the $N_i$, is an estimate of the density $f_X$.

## 8.2.2 Example: BMI Data

Consider the Pima diabetes study from Section 7.8. One of the columns is Body Mass Index (BMI). Let's plot a histogram:

```
pima <-
   read.csv('pima-indians-diabetes.data',
      header=FALSE)
bmi <- pima[,6]
bmi <- bmi[bmi > 0]
hist(bmi,breaks=20,freq=FALSE)
```

The plot is shown in Figure 8.1.

What does the above code do? First, the data must be cleaned. This dataset is well known to have some missing values, coded as 0s, e.g., in BMI and blood pressure.

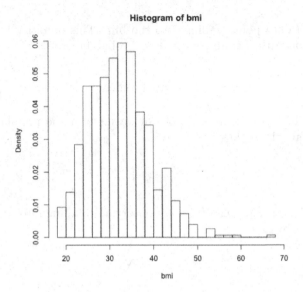

Figure 8.1: BMI, 20 bins

Now, what about the call to **hist()** itself? The **breaks** argument sets the number of bins. I chose 20 here.

Normally the vertical axis in a histogram measures bin counts, i.e., frequencies of occurrence. Setting **freq = FALSE** specifies that we instead want our plot to have area 1.0, as densities do. We thus divide the bin counts by $wn$, so that we get a density estimate.

### 8.2.3   The Number of Bins

Why is there an issue anyway? Here is the intuition:

- If we use too many bins, the graph will be quite choppy. Figure 8.2 shows a histogram for the BMI data with 100 bins. Presumably the true population density is pretty smooth, so the choppiness is a problem.

- On the other hand, if we use too few bins, each bin will be very wide, so we won't get a very detailed estimate of the underlying density. In the extreme, with just *one* bin, the graph becomes completely uninformative.

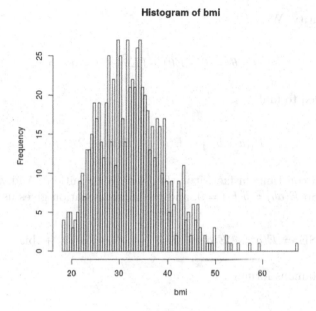

**Histogram of bmi**

Figure 8.2: BMI, 100 bins

### 8.2.3.1 The Bias-Variance Tradeoff

It's instructive to think of the issue of choosing the number of bins in terms of variance and bias, the famous *bias-variance tradeoff*. This is a fundamental issue in statistics. We'll discuss it here in the context of density estimation, but it will return as a core point in Chapter 15.

Suppose we wish to estimate some population quantity $\theta$, using an estimator $\widehat{\theta}$ computed from our sample data. Then we hope to keep the mean squared error,

$$\text{MSE} = E[(\widehat{\theta} - \theta)^2] \tag{8.5}$$

as small as possible. (Once again, keep in mind that $\widehat{\theta}$ is a random variable; each random sample from this population will yield a different value of $\widehat{\theta}$. $\theta$ on the other hand is a fixed, though unknown, quantity.) Let's expand that quantity. Write

$$\widehat{\theta} - \theta = (\widehat{\theta} - E\widehat{\theta}) + (E\widehat{\theta} - \theta) = a + b \qquad (8.6)$$

So, we need to find

$$E[(a + b)^2] = E(a^2) + E(b^2) + 2E(ab) \qquad (8.7)$$

But $b$ is a constant; in fact, it is the bias, by definition. And $Ea = E\widehat{\theta} - E\widehat{\theta} = 0$. So $E(ab) = b \, Ea = 0$, and the above equation gives us

$$\text{MSE} = E(a^2) + E(b^2) = E(a^2) + b^2 = Var(\widehat{\theta}) + \text{bias}^2 \qquad (8.8)$$

That's a famous formula:

$$\text{MSE} = \text{variance plus squared bias} \qquad (8.9)$$

And it is called a *tradeoff* because those two terms are often at odds with each other. For instance, recall the discussion in Section 7.6. The classic definition of the sample variance uses a divisor of $n - 1$, while the one in this book ("our" definition) uses $n$. As was pointed out in that section, the difference is usually minuscule, but this will illustrate the "tradeoff" issue.

- The classic estimator has 0 bias, whereas our bias is nonzero. So, the classic estimator is better in that its second term in (8.9) is smaller.

- On the other hand, since $1/(n - 1) > 1/n$, the classic estimator has a larger variance, a factor of $[n/(n - 1)]^2$ larger (from (4.12)). Thus our estimator has a smaller first term in (8.9).

The overall "winner" will depend on $n$ and the size of $Var(s^2)$. Calculating the latter would be too much of a digression here, but the point is that there IS a tradeoff.

### 8.2.3.2   The Bias-Variance Tradeoff in the Histogram Case

Let's look at the bin width issue in the context of variance and bias.

(a) If the bins are too narrow, then for a given bin, there will be a lot of variation in height of that bin from one sample to another — i.e., the variance of the height will be large.

(b) On the other hand, making the bins too wide produces a bias problem. Say for instance the true density $f_X(t)$ is increasing in $t$, as for instance in the example in Section 6.6. Then within a bin, our estimate $\widehat{f}_X(t)$ will tend to be too low near the left end of the bin and too high on the right end. If the number of bins is small, then the bin widths will be large, and bias may be a serious issue.

Here is another way to see (a) above. As noted, $N_i$ has a binomial distribution with parameters $n$ and $p$. Thus

$$Var(N_i) = np(1 - p) \tag{8.10}$$

and so

$$Var(\widehat{f}_X(c)) = \frac{1}{w^2} \cdot \frac{p(1 - p)}{n} \tag{8.11}$$

Also, again by the binomial property,

$$E[\widehat{f}_X(c)] = \frac{1}{wn} \cdot np = \frac{p}{w} \tag{8.12}$$

Now here is the key point: Recall Section 4.1.3, titled "Intuition Regarding the Size of Var(X)." It was pointed out there that one way to gauge whether a variance is large is to compute the *coefficient of variation*, i.e., ratio of the standard deviation to the mean. In our case here, that is the ratio of the square root of (8.11) to (8.12):

$$\frac{\sqrt{np(1 - p)}}{np} = \frac{1}{\sqrt{n}} \sqrt{\frac{1 - p}{p}} \tag{8.13}$$

Now, for fixed $n$, if we use too many bins, the bin width will be very narrow, so the value of $p$ will be near 0. That would make the coefficient of variation (8.13) large. So here is mathematical confirmation of the qualitative description in bullet (a) above.

But...if $n$ is large, variance is less problematic; a large value of the second factor in (8.13) can be compensated with the small value of the first factor.

So we can afford to make the bin size narrower, thus avoiding excessive bias, pointed out in bullet (b).

In other words, **the larger our sample, the more bins we should have**.

That still begs the question of how *many* bins, but at least we can explore that number armed with this insight. More in the next section.

### 8.2.3.3   A General Issue: Choosing the Degree of Smoothing

Recall the quote in the Preface of this book, from the ancient Chinese philosopher Confucius:

> *[In spite of] innumerable twists and turns, the Yellow River flows east.*

Confucius' point was basically that one should, as we might put it today, "Look at the big picture," focusing on the general eastward trend of the river, rather than the local kinks. We should visually "smooth" our image of the river.

That's exactly what a histogram does. The fewer the number of bins, the more smoothing is done. So, choosing the number of bins can be described as choosing the amount of smoothing. This is a central issue in statistics and machine learning, and will play a big role in Chapter 15 as well as here.

There are various methods for automatic selection of the number of bins [39]. They are too complex to discuss here, but the R package **histogram** [32] offers several such methods. Here is the package in action on the BMI data:

```
histogram(bmi,type='regular')
```

The second argument specifies that we want all bin widths to be the same. The plot is shown in Figure 8.3.

Note that 14 bins were chosen. The graph looks reasonable here, but the reader should generally be a bit wary of automatic methods to select a degree of smoothing, both here and generally. There is no perfect method, and different methods will give somewhat different results.

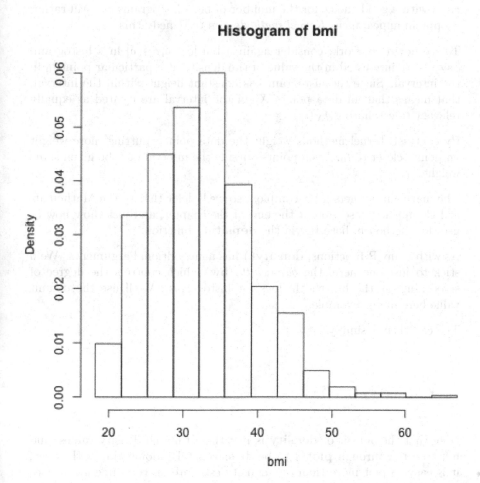

Figure 8.3: BMI, **histogram** package fit

## 8.3    Advanced Methods for Model-Free Density Estimation

Even with a good choice for the number of bins, histograms are still rather choppy in appearance. *Kernel methods* aim to remedy this.

To see how they work, consider again a bin $[c - \delta, c + \delta]$ in a histogram. Say we are interested in the value of the density at a particular point $t_0$ in the interval. Since the histogram has constant height within the interval, that means that all data points $X_i$ in the interval are treated as equally relevant to estimating $f_X(t_0)$.

By contrast, kernel methods weight the data points, putting more weight on points closer to $t_0$. Even points outside the interval may be given some weight.

The mathematics gets a bit complex, so we'll defer that to the Mathematical Complements section at the end of the chapter, and just show how to use this method in base R, via the **density()** function.

As with many R functions, **density()** has many optional arguments. We'll stick to just one here, the *bandwidth*, **bw**, which controls the degree of smoothing, as the bin width does for histograms. We'll use the default value here in our example.[2]

The call then is simply

```
plot(density(bmi))
```

Note that the output of **density** is just the estimated density values, and must be run through **plot()** to be displayed.[3] By doing things this way, it is easy to plot more than one density estimate on the same graph (see Exercise 8).

The graph is shown in Figure 8.4.

---

[2]The default value is an "optimal" one generated by an advanced mathematical argument, not shown here. The reader still should experiment with different bandwidth values, though.

[3]This also involves R *generic* functions. See the Computational Complements at the end of this chapter.

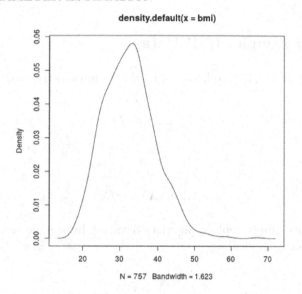

Figure 8.4: Kernel density estimate, BMI data

## 8.4   Parameter Estimation

To fit a parametric model such as the gamma to our data, the question then arises as to how to estimate the parameters. Two common methods for estimating the parameters of a density are the Method of Moments (MM) and Maximum Likelihood Estimation (MLE).[4] We'll introduce these via examples.

### 8.4.1   Method of Moments

MM gets its name from the fact that quantities like mean and variance are called *moments*. $E(X^k)$ is the $k^{th}$ moment of $X$, with $E[(X - EX)^k]$ being termed the $k^{th}$ *central* moment. if we have an $m$-parameter family, we "match" $m$ moments, as follows.

---

[4]These methods are used more generally as well, not just for estimating density parameters.

## 8.4.2    Example: BMI Data

From Section 6.7.4.1, we know that for a gamma-distributed $X$,

$$EX = r/\lambda \qquad (8.14)$$

and

$$Var(X) = r/\lambda^2 \qquad (8.15)$$

In MM, we simply replace population values by sample estimates in the above equations, yielding

$$\overline{X} = \widehat{r}/\widehat{\lambda} \qquad (8.16)$$

and

$$s^2 = \widehat{r}/\widehat{\lambda}^2 \qquad (8.17)$$

Dividing the first equation by the second, we obtain

$$\widehat{\lambda} = \overline{X}/s^2 \qquad (8.18)$$

and thus from the first equation,

$$\widehat{r} = \overline{X}\widehat{\lambda} = \overline{X}^2/s^2 \qquad (8.19)$$

Let's see how well the model fits, at least visually:

```
xb <- mean(bmi)
s2 <- var(bmi)
lh <- xb/s2
ch <- xb^2/s2
hist(bmi,freq=FALSE,breaks=20)
curve(dgamma(x,ch,lh),0,70,add=TRUE)
```

The plot is shown in Figure 8.5. Visually, the fit looks fairly good. Be sure to keep in mind the possible sources of discrepancy between the fitted model and the histogram:

- Sampling variation: We are of course working with sample data, not the population. It may be that with a larger sample, the discrepancy may be lesser.

- Model bias: As the quote from George Box reminds us, a model is just that, a simplifying model of reality. Most models are imperfect, e.g., the assumed massless, frictionless string from physics computations, but are often good enough for our purposes.

- Choice of number of bins: The parametric model here might fit even better with a different choice than our 20 for the number of bins.

### 8.4.3 The Method of Maximum Likelihood

To see how MLE works, consider the following game. I toss a coin until I accumulate $r$ heads. I don't tell you what value I've chosen for $r$, but I do tell you $K$, the number of tosses I needed. You then must guess the value of $r$. Well, $K$ has a negative binomial distribution (Section 5.4.3), so

$$P(K = u) = \binom{u-1}{r-1} 0.5^u, \ u = r, r+1, ... \tag{8.20}$$

Say I tell you $K = 7$. Then what you might do is find the value of $r$ that maximizes

$$\binom{6}{r-1} 0.5^7 \tag{8.21}$$

You are asking, "What value of $r$ would have made our data ($K = 7$) most likely?" Trying $r = 1, 2, ..., 7$, one finds that $r = 4$ maximizes (8.21), so we would take that as our guess.[5]

Now consider our parametric density setting. For "likelihood" with continous data, we don't have probabilities, but it is defined in terms of densities, as follows.

---

[5]By the way, here is how the Method of Moments approach would work here. For the negative binomial distribution it is known that $E(K) = r/p$, where $p$ is the probability of "success," in this setting meaning heads. So $E(K) = 2r$. Under MM, we would set $\overline{K} = 2\hat{r}$, where the left-hand side is the average of all values of $K$ in our data. We only did the "experiment" once, so $\overline{K} = 6$ and we guess $r$ to be 3.

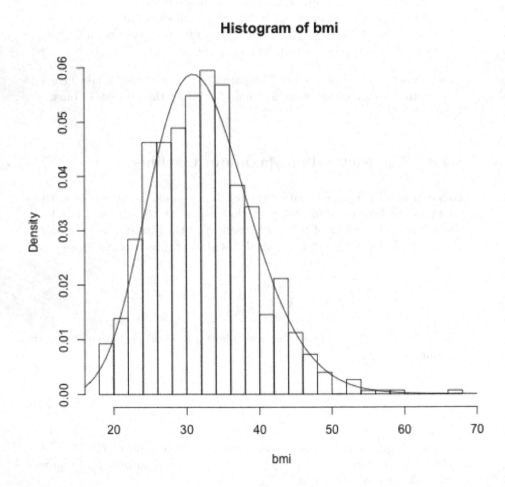

Figure 8.5: BMI, histogram and gamma fit

Say $g(t, \theta)$ is our parametric density, with $\theta$ being the parameter (possibly vector-valued). The likelihood is defined as

$$L = \Pi_i^n g(X_i, \theta) \qquad (8.22)$$

We will take $\widehat{\theta}$ to be the value that maximizes $L$, but it's usually easier to equivalently maximize

$$l = \log L = \Sigma_i^n \log g(X_i, \theta) \qquad (8.23)$$

Typically the equations have no closed-form solution, and thus must be solved numerically. R's **mle()** function does this for us.

### 8.4.4   Example: Humidity Data

This is from the Bike Sharing dataset on the UC Irvine Machine Learning Repository [12]. We are using the **day** data, one column of which is for humidity.

Since the humidity values are in the interval (0,1), a natural candidate for a parametric model would be a beta distribution (Section 6.7.5). Here is the code and output:

```
> bike <- read.csv('day.csv',header=TRUE)
> hum <- bike$hum
> hum <- hum[hum > 0]
> library(stats4)
> ll <- function(a,b)
+    sum(-log(dbeta(hum,a,b)))
> z <- mle(minuslogl=ll,start=list(a=1,b=1))
> z
...
Coefficients:
        a          b
6.439144 3.769841
```

The R function **mle()** has two main arguments. The first specifies a function that computes the log-likelihood, our function **ll()** here. The arguments to that function must be the parameters, which I have named **a** and **b** for "alpha" and "beta."

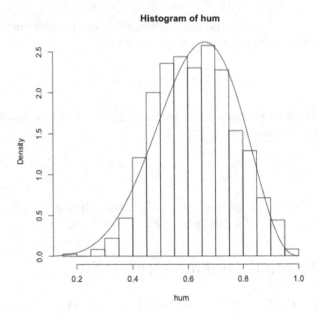

Figure 8.6: Humidity, histogram + fitted beta density

The calculation uses an iterative process, starting with an initial guess for the MLEs, then successsively refining the guess until convergence to the MLEs. The second argument to **mle()**, **start**, specifies our initial guess.

Let's plot this fitted density against the histogram:

```
> hist(hum,breaks=20,freq=FALSE)
> a <- coef(z)[1]
> b <- coef(z)[2]
> curve(dbeta(x,a,b),0,1,add=TRUE)
```

The result is shown in Figure 8.6. The caveats at the end of Section 8.4.2 apply here as well.

By the way, **mle()** also provides standard errors for the estimates $\alpha$ and $\beta$:

```
> vcov(z)
           a          b
a 0.11150334 0.05719833
b 0.05719833 0.03616982
```

This is the *covariance* matrix, with variances on the diagonal and covariances in the off-diagonal slots. So the standard error of $\widehat{\alpha}$ is $\sqrt{0.11150334}$, or about 0.334.

## 8.5   MM vs. MLE

MM and MLE are both powerful techniques, but which is better? On the one hand, MLEs can be shown to asymptotically optimal (smallest standard errors). On the other hand, MLEs require more assumptions. As with many things in data science, the best tool may depend on the setting.

## 8.6   Assessment of Goodness of Fit

In our examples above, we can do a visual assessment of how well our model fits the data, but it would be nice to have a quantitative measure of goodness of fit.

The classic assessment tool is the Chi-Squared Goodness of Fit Test. It is one of the oldest statistical methods (1900!), and thus in wide use. But Professor Box's remark suggests that this procedure is not the best way to gauge model fit, as the test answers the yes–or–no question, e.g., "Is the population distribution *exactly* gamma?" — of dubious relevance, given that we know *a priori* that the answer is No.[6]

A more useful measure is the *Kolmogorov-Smirnov (KS) statistic*. It actually gives us the size of the discrepancy between the fitted model family and the true population distribution. To make matters concrete, say we are fitting a beta model, with the humidity data above..

K-S is based on cdfs. Of course, the **pbeta()** function gives us the cdf for the beta family, but we also need a model-free estimate of $F_X$, the true population cdf of $X$. For the latter, we use the *empirical cdf* of $X$, defined as

$$\widehat{F}_X(t) = \frac{M(t)}{n} \tag{8.24}$$

where $M(t)$ is simply a count of the number of $X_i$ that are $\leq t$. The R function **ecdf()** calculates this for us:

---

[6]This is a problem with significance tests in general, to be discussed in Chapter 10.

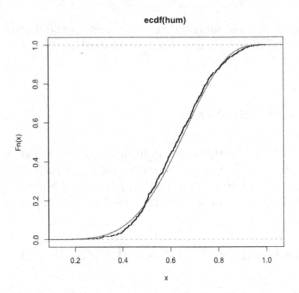

Figure 8.7: K-S analysis, humidity data

```
> ehum <- ecdf(hum)
> plot(ehum,cex=0.1)
> curve(pbeta(x,a,b),0,1,add=TRUE)
```

(The values of **a** and **b** had been previously computed.)  The plot is in
Figure 8.7.

So that's the visual, showing a good fit.  Now to actually quantify the
situation, the K-S statistic measures the fit, in terms of the maximum
discrepancy, i.e., the largest difference between the empirical cdf and the
fitted cdf:

```
> fitted.beta <- pbeta(hum,a,b)
> eh <- ecdf(hum)
> ks.test(eh,fitted.beta)$statistic
          D
0.04520548
...
```

Since cdf values range in [0,1], that maximum discrepancy of 0.045 is pretty
good.

Of course, we mustn't forget that this number is subject to sampling variation. We can partially account for that with a K-S *confidence band* [20].[7]

# 8.7 The Bayesian Philosophy

*Everyone is entitled to his own opinion, but not his own facts* — Daniel Patrick Moynihan, senator from New York, 1976-2000

*Black cat, white cat, it doesn't matter as long as it catches mice* — Deng Xiaoping, when asked about his plans to give private industry a greater role in China's economy

*Whiskey's for drinkin' and water's for fightin' over* — Mark Twain, on California water jurisdiction battles

Over the years, the most controversial topic in statistics by far has been that of *Bayesian* methods, the "California water" of the statistics world. Though usage is common today, the topic remains one for which strong opinions are held on both sides.

The name stems from Bayes' Rule (Section 1.9),

$$P(A|B) = \frac{P(A)P(B|A)}{P(A)P(B|A) + P(\text{not } A)P(B|\text{not } A)} \qquad (8.25)$$

No one questions the validity of Bayes' Rule, and thus there is no controversy regarding statistical procedures that make use of probability calculations based on that rule. But the key word is *probability*. As long as the various terms in (8.25) are real probabilities — that is, based on actual data — there is no controversy.

But instead, the debate is over the cases in which Bayesians replace some of the probabilities in the theorem with "feelings," i.e., non-probabilities, arising from what they call *subjective prior distributions*. The key word is then *subjective*.

By contrast, there is no controversy if the prior makes use of real data, termed *empirical Bayes*. Actually, many Bayesian analyses one sees in practice are of this kind, and again, thre is no controversy here. So, our use of the term here *Bayesian* refers only to subjective priors.

---

[7]As the name implies, **ks.test()** also offers a signficance test, but we do not use it here for the same reasons given above regarding the chi-squared test.

Say we wish to estimate a population mean. Here the Bayesian analyst, before even collecting data, says, "Well, I think the population mean could be 1.2, with probability, oh, let's say 0.28, but on the other hand, it might also be 0.88, with probability, well, I'll put it at 0.49..." etc. This is the analyst's subjective prior distribution for the population mean. The analyst does this before even collecting any data. Note carefully that he is NOT claiming these are real probabilities; he's just trying to quantify his hunches. The analyst then collects the data, and uses some mathematical procedure that combines these "feelings" with the actual data, and which then outputs an estimate of the population mean or other quantity of interest.

### 8.7.1   How Does It Work?

The technical details can become quite involved. The reader is referred to [10] for an in-depth treatment by leaders in the field, but we can at least introduce the methodology here.

Say our data is assumed to have a Poisson distribution, and we wish to etimate $\lambda$. Keep in mind that the latter is now treated as a random variable, and we are interested in finding the *posterior distribution*, $f_{\lambda|\text{ the }X_i}$. Then (8.25) would be something like

$$f_{\lambda|\text{ the }X_i} = \frac{f_\lambda \, p_{\text{ the }X_i|\lambda}}{\int p_{\text{ the }X_i|\lambda} \, f_\lambda \, d\lambda} \qquad (8.26)$$

For our distribution for $\lambda$, we might choose a *conjugate prior*, meaning one for which (8.26) has a convenient closed form. For the Poisson case, a conjugate prior turns out to be a gamma distribution. The analyst applies her feeling about the setting by choosing the parameters of the latter accordingly. Then (8.26) turns out also to be gamma. We could then take our estimate of $\lambda$ to be, say, the conditional mode.

### 8.7.2   Arguments For and Against

The Bayesians justify this by saying one should use all available information, even if it is just a hunch. "The analyst is typically an expert in the field under study. You wouldn't want to throw away his/her expertise, would you?" Moreover, they cite theoretical analyses that show that Bayes estimator doing very well in terms of criteria such as mean squared error, even if the priors are not "valid."

The non-Bayesians, known as *frequentists*, on the other hand dismiss this as unscientific and lacking in impartiality. "In research on a controversial health issue, say, you wouldn't want the researcher to incorporate his/her personal political biases into the number crunching, would you?" So, the frequentists' view is reminiscent of the Moynihan quote above.

In the computer science/machine learning world Bayesian estimation seems to be much less of a controversy.[8] Computer scientists, being engineers, tend to be interested in whether a method seems to work, with the reasons being less important. This is the "black cat, white cat" approach in the Deng quote above.

## 8.8   Mathematical Complements

### 8.8.1   Details of Kernel Density Estimators

How does the kernel method (Section 8.3) work? Recall that this method for estimating $f_X(t)$ is similar to a histogram, but gives heavier weight to data points near $t$.

One chooses a weight function $k()$, the *kernel*, which can be any nonnegative function integrating to 1.0. (This of course is a density in its own right, but it is just playing the role of a weighting scheme and is not related to the density being estimated.) The density estimate is then

$$\widehat{f}_X(t) = \frac{1}{nh} \sum_{i=1}^{n} k\left(\frac{t - X_i}{h}\right) \tag{8.27}$$

where $h$ is the bandwidth.

Say for instance one chooses to use a N(0,1) kernel. For an $X_i$ very near $t$ (in units of $h$), the quantity $(t - X_i)/h$ will be near 0, the peak of $k()$; thus this particular $X_i$ will be given large weight. For an $X_j$ far from $t$, the weight will be small.

As noted, the R function **density()** performs these calculations.

---

[8]Note again, though, that in many cases they are using empirical Bayes, not subjective.

# 8.9    Computational Complements

## 8.9.1    Generic Functions

In Section 8.3, we saw that the output of **density()** is not automatically plotted; we need to call **plot()** on it for display. Let's take a closer look:

```
> z <- density(bmi)
> class(z)
[1] "density"
> str(z)
List of 7
 $ x         : num [1:512] 13.3 13.4 13.6 13.7 ...
 $ y         : num [1:512] 1.62e-05 2.02e-05 ...
 $ bw        : num 1.62
 $ n         : int 757
 $ call      : language density.default(x = bmi)
 $ data.name : chr "bmi"
 $ has.na    : logi FALSE
 - attr(*, "class")= chr "density"
```

(R's **str()** function shows a summary of the "innards" of an object.)

So, the output is an object of R's S3 class structure, containing the horizontal and vertical coordinates of the points to be plotted, the bandwidth and so on. Most important, the class is **'density'**. Here is how that is used.

The *generic* R function **plot()** is just a placeholder, not an actual function. (Section A.9.2.) When we call it, R checks the class of the argument, then *dispatches* the call to a class-specific plotting function, in this case **plot.density()**.

To see this in action, let's run R's **debug()** function:[9]

```
> debug(plot)
> plot(z)
debugging in: plot(z)
debug: UseMethod("plot")
Browse[2]> s
debugging in: plot.density(z)
debug: {
```

---

[9]In the software world, a debugging tool is often useful as a way to get to understand code, not just for debugging.

```
    if (is.null(xlab))
...
```

Sure enough, we're in **plot.density()**.

R has many other *generic* functions besides **plot()**, such as **print()**, **summary()** and notably in Chapter 15, **predict()**.

## 8.9.2   The gmm Package

GMM, the Generalized Method of Moments, was developed by Lars Peter Hansen, who shared the 2013 Nobel Prize in Economics in part for this [19]. As the name implies, its scope is far broader than ordinary MM, but we will not pursue that here.

### 8.9.2.1   The gmm() Function

Earlier we saw **mle()**, a function in base R for numerically solving for MLEs in settings where the MLEs don't have closed-form solutions. The **gmm** package [7] does this for MM estimation.

The **gmm()** function in the package is quite simple to use. As with **mle()**, it is an iterative method. The form we'll use is

```
gmm(g,x,t0)
```

Here **g** is a function to calculate the moments (analgous to the funciton **ll** in **mle**, which calculates the likelihood function); **x** is our data; and **t0** is a vector of initial guesses for the parameters.

The estimated covariance matrix of $\widehat{\theta}$ can be obtained by calling **vcov()** on the object returned by **gmm()**, thus enabling the calculation of standard errors.

### 8.9.2.2   Example: Bodyfat Data

Our dataset will be **bodyfat**, which is included in the **mfp** package [36], with measurements on 252 men. The first column of this data frame, **brozek**, is the percentage of body fat, which when converted to a proportion is in (0,1). That makes it a candidate for fitting a beta distribution.

```
> library(mfp)
```

```
> data(bodyfat)
> g <- function(th,x) {
+       t1 <- th[1]
+       t2 <- th[2]
+       t12 <- t1 + t2
+       meanb <- t1 / t12
+       m1 <- meanb - x
+       m2 <- t1*t2 / (t12^2 * (t12+1)) - (x-meanb)^2
+       cbind(m1,m2)
+ }
> gmm(g,bodyfat$brozek/100,c(alpha=0.1,beta=0.1))
...
   alpha       beta
  4.6714   19.9969
...
```

As mentioned, **g()** is a user-supplied function that calculates the moments. It depends on **th**, our latest iterate in finding $\widehat{th}$, and our data **x**. Our function here calculates and returns the first two moments, **m1** and **m2**, according to (6.53) and (6.54).

## 8.10   Exercises

**Mathematical problems:**

1.  Suppose the distribution of some random variable $X$ is modeled as uniform on $(0, c)$. Find a closed-form expression for the MM estimator of $c$, based on data $X_1, ..., X_n$.

2.  Suppose the distribution of some random variable $X$ is modeled as uniform on $(r, s)$. Find a closed-form expression for the MM estimator of $r$ and $s$, based on data $X_1, ..., X_n$.

3.  Consider the parametric density family $ct^{c-1}$ for $t$ in (0,1), 0 elsewhere. Find closed-form expressions for the MLE and the MM estimate, based on data $X_1, ..., X_n$.

4.  Suppose in a certain population $X$ has an exponential distribution with parameter $\lambda$. Consider a histogram bin $[c - \delta, c + \delta]$. Derive a formula, in

terms of $\lambda$, $c$, $\delta$ and the sample size $n$, for $E[\widehat{f}_X(c)]$.

**5.** Find expressions involving $F_X$ for the mean and variance of $\widehat{F}_X(t)$ in (8.24).

## Computational and data problems

**6.** Find the MLE for the BMI data, and compare to MM.

**7.** Plot a histogram of the **Insul** column in the Pima data, and fit a parameter density model of your choice to it.

**8.** For the BMI data, plot two density estimates on the same graph, using **plot()** for the first then **lines()** for the second.

**9.** Use **gmm()** to fit a beta model to the insulin data.

**10.** Continue the analysis in Section 8.9.2.2, by calculating the K-S goodness-of-fit measure.

**11.** Use **mle()** to find the MLE of $\gamma$ for a power law in the e-mail data, Section 5.5.3.3.

**12.** On page 183, the possibility is raised that the gamma fit may seem more accurate if we change the number of histogram bins from 20 to something else. Try this, using several bin numbers of your choice.

**13.** Suppose we observe $X$, which is binomially distributed with success probability 0.5. The number of trials $N$ is assumed to be geometrically distributed with success probability also 0.5. Since the number of trials in a binomial distribution is a parameter of the binomial family, this setting could be viewed as Bayesian, with a geometric prior for $N$, the number of trials. Write a function with call form **g(k)** that returns the conditional mode (i.e., mostly likely value) of $N$, given $X = k$.

# Chapter 9

# The Family of Normal Distributions

Again, these are the famous "bell-shaped curves," so called because their densities have that shape.

## 9.1   Density and Properties

The density for a normal distribution is

$$f_W(t) = \frac{1}{\sqrt{2\pi}\sigma} \, e^{-0.5\left(\frac{t-\mu}{\sigma}\right)^2}, \; -\infty < t < \infty \qquad (9.1)$$

Again, this is a two-parameter family, indexed by the parameters $\mu$ and $\sigma$, which turn out to be the mean[1] and standard deviation, $\mu$ and $\sigma$. The notation for it is $N(\mu, \sigma^2)$ (it is customary to state the variance $\sigma^2$ rather than the standard deviation).

And we write

$$X \sim N(\mu, \sigma^2) \qquad (9.2)$$

to mean that the random variable $X$ has the distribution $N(\mu, \sigma^2)$. (The tilde is read "is distributed as.")

---

[1]Remember, this is a synonym for expected value.

## 9.1.1   Closure under Affine Transformation

The family is closed under affine transformations:

If

$$X \sim N(\mu, \sigma^2) \tag{9.3}$$

and for constants $c$ $d$ we set

$$Y = cX + d \tag{9.4}$$

then

$$Y \sim N(c\mu + d, c^2\sigma^2) \tag{9.5}$$

For instance, suppose $X$ is the height of a randomly selected UC Davis student, measured in inches. Human heights do have approximate normal distributions; a histogram plot of the student heights would look bell-shaped. Now let $Y$ be the student's height in centimeters. Then we have the situation above, with $c = 2.54$ and $d = 0$. The claim about affine transformations of normally distributed random variables would imply that a histogram of $Y$ would again be bell-shaped.

Consider the above statement carefully.

The statement is saying much more than simply that $Y$ has mean $c\mu + d$ and variance $c^2\sigma^2$, which would follow from our our "mailing tubes" such as (4.12) *even if $X$ did not have a normal distribution.* The key point is that this new variable $Y$ is also a member of the normal family, i.e., its density is still given by (9.1), now with the new mean and variance.

Let's derive this.[2] For convenience, suppose $c > 0$. Then

---

[2] The reader is asked to be patient here! The derivation is a bit long, but it will serve to solidify various concepts in the reader's mind.

$$
\begin{aligned}
F_Y(t) &= P(Y \le t) \quad \text{(definition of } F_Y) & (9.6)\\
&= P(cX + d \le t) \quad \text{(definition of Y)} & (9.7)\\
&= P\left(X \le \frac{t-d}{c}\right) \quad \text{(algebra)} & (9.8)\\
&= F_X\left(\frac{t-d}{c}\right) \quad \text{(definition of } F_X) & (9.9)
\end{aligned}
$$

Therefore

$$
\begin{aligned}
f_Y(t) &= \frac{d}{dt}F_Y(t) \quad \text{(definition of } f_Y)\\
&= \frac{d}{dt}F_X\left(\frac{t-d}{c}\right) \quad \text{(from (9.9))}\\
&= f_X\left(\frac{t-d}{c}\right) \cdot \frac{d}{dt}\frac{t-d}{c} \quad \text{(definition of } f_X \text{ and Chain Rule)}\\
&= \frac{1}{c} \cdot \frac{1}{\sqrt{2\pi}\sigma} e^{-0.5\left(\frac{\frac{t-d}{c}-\mu}{\sigma}\right)^2} \quad \text{(from (9.1)}\\
&= \frac{1}{\sqrt{2\pi}(c\sigma)} e^{-0.5\left(\frac{t-(c\mu+d)}{c\sigma}\right)^2} \quad \text{(algebra)}
\end{aligned}
$$

That last expression is the $N(c\mu + d, c^2\sigma^2)$ density, so we are done!

## 9.1.2 Closure under Independent Summation

If $X$ and $Y$ are independent random variables, each having a normal distribution, then their sum $S = X + Y$ also is normally distributed.

*This is a pretty remarkable phenomenon!* It is not true for most other parametric families. If for instance $X$ and $Y$ each with, say, a $U(0,1)$ distribution, then the density of $S$ turns out to be triangle-shaped, NOT another uniform distribution. (This can be derived using the methods of Section 11.8.1.)

Note that if $X$ and $Y$ are independent and normally distributed, then the two properties above imply that $cX + dY$ will also have a normal distribution, for any constants $c$ and $d$.

More generally:

For constants $a_1, ..., a_k$ and *independent* random variables $X_1, ..., X_k$, with

$$X_i \sim N(\mu_i, \sigma_i^2) \tag{9.10}$$

form the new random variable $Y = a_1 X_1 + ... + a_k X_k$. Then

$$Y \sim N(\sum_{i=1}^{k} a_i \mu_i, \sum_{i=1}^{k} a_i^2 \sigma_i^2) \tag{9.11}$$

### 9.1.3  A Mystery

Again, the reader should ponder how remarkable this property — the sum of two independent normal random variables is again normal — of the normal family is, because it would appear not to have an intuitive explanation.

Imagine random variables $X$ and $Y$, each with a normal distribution. Say the mean and variances are 10 and 4 for $X$, and 18 and 6 for $Y$. We repeat our experiment 1000 times for our "notebook," i.e., 1000 lines with two columns. If we draw a histogram of the $X$ column, we'll get a bell-shaped curve, and the same will be true for the $Y$ column.

But now add a $Z$ column, for $Z = X + Y$. Why in the world should a histogram of the $Z$ column also be bell-shaped? (We'll revisit this later.)

## 9.2  R Functions

```
dnorm(x, mean = 0, sd = 1)
pnorm(q, mean = 0, sd = 1)
qnorm(p, mean = 0, sd = 1)
rnorm(n, mean = 0, sd = 1)
```

Here **mean** and **sd** are of course the mean and standard deviation of the distribution. The other arguments are as in our previous examples.

## 9.3  The Standard Normal Distribution

**Definition 17** *If* $Z \sim N(0, 1)$ *we say the random variable* $Z$ *has a standard normal distribution.*

*Note that if $X \sim N(\mu, \sigma^2)$, and if we set*

$$Z = \frac{X - \mu}{\sigma} \qquad (9.12)$$

*then*

$$Z \sim N(0, 1) \qquad (9.13)$$

The above statements follow from the earlier material:

- Define $Z = \frac{X - \mu}{\sigma}$.

- Rewrite it as $Z = \frac{1}{\sigma} \cdot X + (\frac{-\mu}{\sigma})$.

- Since $E(cU + d) = cEU + d$ for any random variable U and constants $c$ and $d$, we have

$$EZ = \frac{1}{\sigma} EX - \frac{\mu}{\sigma} = 0 \qquad (9.14)$$

  and (4.19) and (4.12) imply that $Var(X) = 1$.

- OK, so we know that $Z$ has mean 0 and variance 1. But *does it have a normal distribution?* Yes, due to our discussion above titled "Closure Under Affine Transformations."

By the way, the $N(0, 1)$ cdf is traditionally denoted by $\Phi$.

## 9.4 Evaluating Normal cdfs

Traditionally, statistics books have included as an appendix a table of the $N(0, 1)$ cdf, formed by numerical approximation methods. This was necessary because the function in (9.1) does not have a closed-form indefinite integral.

But this raises a question: There are infinitely many distributions in the normal family. Don't we need a separate table for each? That of course would not be possible, and in fact it turns out that this one table—the one for the $N(0, 1)$ distribution— is sufficient for the entire normal family.

Though we of course will use R to get such probabilities, it will be quite instructive to see how these table operations work.

The key is the material in Section 9.3 above. Say $X$ has an $N(10, 2.5^2)$ distribution. How can we get a probability like, say, $P(X < 12)$ using the $N(0, 1)$ table? Write

$$P(X < 12) = P\left(Z < \frac{12 - 10}{2.5}\right) = P(Z < 0.8) \qquad (9.15)$$

Since on the right-hand side $Z$ has a standard normal distribution, we can find that latter probability from the $N(0, 1)$ table!

In the R statistical package, the normal cdf for any mean and variance is available via the function **pnorm()**. In the above example, we just run

```
> pnorm(12,10,2.5)
[1] 0.7881446
```

## 9.5   Example: Network Intrusion

As an example, let's look at a simple version of the network intrusion problem, a major aspect of computer security. Suppose we have found that in Jill's remote logins to a certain computer, the number $X$ of disk sectors she reads or writes has an approximate normal distribution with a mean of 500 and a standard deviation of 15.

Before we continue, a comment on modeling: Since the number of sectors is discrete, it could not have an exact normal distribution. But then, no random variable in practice has an exact normal or other continuous distribution, as discussed in Section 6.2, but the distribution can indeed by approximately normal.

Now, say our network intrusion monitor finds that Jill—or someone posing as her—has logged in and has read or written 535 sectors. Should we be suspicious? If it really is Jill, how likely would it be for her to read this many sectors or more?

```
> 1 - pnorm(535,500,15)
[1] 0.009815329
```

That 0.01 probability makes us suspicious. While it *could* really be Jill, this would be unusual behavior for Jill, so we start to suspect that it isn't her. It's suspicious enough for us to probe more deeply, e.g., by looking at which files she (or the impostor) accessed — were they rare for Jill too? What about time of day, location from which the access was made, and so on?

Now suppose there are two logins to Jill's account, accessing $X$ and $Y$ sectors, with $X + Y = 1088$. Is this rare for her, i.e., is $P(X + Y > 1088)$? small?

We'll assume $X$ and $Y$ are independent. We'd have to give some thought as to whether this assumption is reasonable, depending on the details of how we observed the logins, etc., but let's move ahead on this basis.

From page 199, we know that the sum $S = X + Y$ is again normally distributed. Due to our mailing tubes on expected value and variance, we know S has mean $2 \cdot 500$ and variance $2 \cdot 15^2 = 450$. The desired probability is then found via

```
1 - pnorm(1088,1000,sqrt(450))
```

which is about 0.00002. That is indeed a small number, and we should be highly suspicious.

Note again that the normal model (or any other continuous model) can only be approximate, especially in the tails of the distribution, in this case far out in the right-hand tail. We shouldn't take the 0.00002 figure too literally. But it is clear that $S$ is only rarely larger than 1088, and the matter mandates further investigation.

Of course, this is very crude analysis, and real intrusion detection systems are much more complex, but you can see the main ideas here.

## 9.6  Example: Class Enrollment Size

After years of experience with a certain course, a university has found that online pre-enrollment in the course is approximately normally distributed, with mean 28.8 and standard deviation 3.1. Suppose that in some particular offering, pre-enrollment was capped at 25, and it hit the cap. Find the probability that the actual demand for the course was at least 30.

Note that this is a conditional probability! Evaluate it as follows. Let $N$ be the actual demand. Then the key point is that we are given that $N \geq 25$, so

$$P(N \geq 30 \mid N \geq 25) \;=\; \frac{P(N \geq 30 \text{ and } N \geq 25)}{P(N \geq 25)} \qquad ((1.8))$$

$$=\; \frac{P(N \geq 30)}{P(N \geq 25)}$$

$$=\; \frac{1 - \Phi\left[(30 - 28.8)/3.1\right]}{1 - \Phi\left[(25 - 28.8)/3.1\right]}$$

$$=\; 0.39$$

Sounds like it may be worth moving the class to a larger room before school starts.

Since we are approximating a discrete random variable by a continuous one, it might be more accurate here to use a *correction for continuity*, described in Section 9.7.2.

## 9.7   The Central Limit Theorem

The Central Limit Theorem (CLT) says, roughly speaking, that a random variable which is a sum of many components will have an approximate normal distribution. So, for instance, human weights are approximately normally distributed, since a person is made of many components. The same is true for SAT raw test scores,[3] as the total score is the sum of scores on the individual problems.

There are many versions of the CLT. The basic one requires that the summands be independent and identically distributed:[4]

**Theorem 18** *Suppose $X_1, X_2, \ldots$ are independent random variables, all having the same distribution which has mean $m$ and variance $v^2$. Form the new random variable $T = X_1 + \ldots + X_n$. Then for large $n$, the distribution of $T$ is approximately normal with mean $nm$ and variance $nv^2$.*

The larger $n$ is, the better the approximation, but typically $n = 25$ is enough.

---

[3]This refers to the raw scores, before scaling by the testing company.

[4]A more mathematically precise statement of the theorem is given in Section 9.11.1.

## 9.7.1   Example: Cumulative Roundoff Error

Suppose that computer roundoff error in computing the square roots of numbers in a certain range is distributed uniformly on (-0.5,0.5), and that we will be computing the sum of $n$ such square roots, say 50 of them. Let's find the approximate probability that the sum is more than 2.0 higher than it should be. (Assume that the error in the summing operation is negligible compared to that of the square root operation.)

Let $U_1, ..., U_{50}$ denote the errors on the individual terms in the sum. Since we are computing a sum, the errors are added too, so our total error is

$$T = U_1 + ... + U_{50} \qquad (9.16)$$

By the Central Limit Theorem, since $T$ is a sum, it has an approximately normal distribution, with mean $50\,EU$ and variance $50\,Var(U)$, where $U$ is a random variable having the distribution of the $U_i$. From Section 6.7.1.1, we know that

$$EU = (-0.5 + 0.5)/2 = 0, \quad Var(U) = \frac{1}{12}[0.5 - (-0.5)]^2 = \frac{1}{12} \qquad (9.17)$$

So, the approximate distribution of $T$ is $N(0, 50/12)$. We can then use R to find our desired probability:

```
> 1 - pnorm(2,mean=0,sd=sqrt(50/12))
[1]  0.1635934
```

## 9.7.2   Example: Coin Tosses

Binomially distributed random variables, though discrete, also are approximately normally distributed. Here's why:

Say $T$ has a binomial distribution with $n$ trials. As we did in Section 5.4.2, write $T$ as a sum of indicator random variables,

$$T = B_1 + ... + B_n \qquad (9.18)$$

where $B_i$ is 1 for a success and 0 for a failure on the $i^{th}$ trial. Since we have a sum of independent, identically distributed terms, the CLT applies.

Thus we use the CLT if we have binomial distributions with large $n$. The mean and variance will be those of the binomial, $np$ and $np(1-p)$.

For example, let's find the approximate probability of getting more than 18 heads in 30 tosses of a coin. The exact answer is

```
> 1 - pbinom(18,30,0.5)
[1] 0.1002442
```

Let's see how close the CLT approximation comes. $X$, the number of heads, has a binomial distribution with $n = 30$ and $p = 0.5$. Its mean and variance are then $np = 15$ and $np(1-p) = 7.5$.

But wait...do we treat this problem as $P(X > 18)$ or $P(X \geq 19)$? If $X$ were a continuous random variable, we wouldn't worry about $>$ versus $\geq$. But $X$ here is discrete, even though are approximating it with a continuous distribution. So let's try both:

```
> 1 - pnorm(18,15,sqrt(7.5))
[1] 0.1366608
> 1 - pnorm(19,15,sqrt(7.5))
[1] 0.07206352
```

Not too surprisingly, one number is too large and the other too small. Why such big errors? The main reason is $n$ here is rather small, but again, the other issue is that we are approximating the distribution of a discrete random variable by a continuous one, which introduces additional error.

But the above numbers suggest that we "split the difference":

```
> 1 - pnorm(18.5,15,sqrt(7.5))
[1] 0.1006213
```

Ah, very nice. This is known as a *correction for continuity*,

### 9.7.3   Example: Museum Demonstration

Many science museums have the following visual demonstration of the CLT.

There are many gumballs in a chute, with a triangular array of $r$ rows of pins beneath the chute. Each gumball falls through the rows of pins, bouncing left and right with probability 0.5 each, eventually being collected into one of $r + 1$ bins, numbered 0 to $r$ from left to right.[5]

---

[5]There are many excellent animations of this on the Web, e.g., the Wikipedia entry for "Bean machine."

A gumball will end up in bin $i$ if it bounces rightward in $i$ of the $r$ rows of pins, Let $X$ denote the bin number at which a gumball ends up. $X$ is the number of rightward bounces ("successes") in $r$ rows ("trials"). Therefore $X$ has a binomial distribution with $n = r$ and $p = 0.5$.

Each bin is wide enough for only one gumball, so the gumballs in a bin will stack up. And since there are many gumballs, the height of the stack in bin $i$ will be approximately proportional to $P(X = i)$. And since the latter will be approximately given by the CLT, the stacks of gumballs will roughly look like the famous bell-shaped curve!

### 9.7.4 A Bit of Insight into the Mystery

Returning to the question raised in Section 9.1.3 — what is the intuition behind the sum $S = X + Y$ of two independent normal random variables itself being normally distributed?

Think of $X$ and $Y$ being approximately normal and arising from the CLT, i.e., $X = X_1 + ... + X_n$ and $Y = Y_1 + ... + Y_n$. Now regroup:

$$S = (X_1 + Y_1) + ...(X_n + Y_n) \tag{9.19}$$

So we see $S$ too would be a sum of i.i.d. terms, so it too would be approximately normal by the CLT.

## 9.8 $\overline{X}$ Is Approximately Normal

The Central Limit Theorem tells us that the numerator in

$$\overline{X} = \frac{X_1 + X_2 + X_3 + ... + X_n}{n} \tag{9.20}$$

from (7.1) has an approximate normal distribution. This has major implications.

### 9.8.1 Approximate Distribution of $X$

:

That means that affine transformations of that numerator are also approximately normally distributed (page 198). Moreover, recall (7.5). Puting these together, we have:

The quantity

$$Z = \frac{\overline{X} - \mu}{\sigma/\sqrt{n}} \qquad (9.21)$$

has an approximately $N(0,1)$ distribution, where $\sigma^2$ is the population variance. This is true regardless of whether the distribution of $X$ in the population is normal.

Remember, we don't know either $\mu$ or $\sigma$; the whole point of taking the random sample is to estimate them. Nevertheless, their values do exist, and thus the fraction $Z$ does exist. And by the CLT, $Z$ will have an approximate $N(0,1)$ distribution.

Make sure you understand why it is the "N" that is approximate here, not the 0 or 1.

So even if the population distribution is very skewed, multimodal and so on, the sample mean will still have an approximate normal distribution. This will turn out to be the core of statistics; they don't call the theorem the *Central* Limit Theorem for nothing!

The reader should make sure he/she fully understands the setting. Our sample data $X_1, ..., X_n$ are random, so $\overline{X}$ is a random variable too. Think in notebook terms: Each row will record a random sample of size $n$. There will be $n + 1$ columns, labeled $X_1, ..., X_n, \overline{X}$.

Say the sampled population has an exponential distribution. Then a histogram of column 1 will look exponential. So will column 2 and so on. But a histogram of column $n + 1$ will look normal!

## 9.8.2   Improved Assessment of Accuracy of $X$

Back in Section 7.7 we used Chebychev's Inequality to assess the accuracy of $\overline{X}$ as an estimator of $\mu$. We found that

In at least 8/9 of all possible samples, $\overline{X}$ is within 3 standard errors, i.e., $3s/\sqrt{n}$, of the population mean $\mu$.

But now, with the CLT, we have

$$P(|\overline{X} - \mu| < 3\sigma/\sqrt{n}) = P(|Z| < 3) \approx 1 - 2\Phi(-3) = 1 - 2 \cdot 0.0027 = 0.9946$$
$$(9.22)$$

Here $Z$ is as in (9.21) and $\Phi()$ is the $N(0,1)$ cdf. The numeric value was obtained by the call **pnorm(-3)**.[6]

Certainly 0.9946 is a more optimistic value than $8/9 = 0.8889$. We in fact are almost certain to have our estimate within 3 standard errors of $\mu$.

## 9.9 Importance in Modeling

Needless to say, there are no random variables in the real world that are exactly normally distributed. In addition to our comments in Section 6.2 that no real-world random variable has a continuous distribution, there are no practical applications in which a random variable is not bounded on both ends. This contrasts with normal distributions, which are continuous and extend from $-\infty$ to $\infty$.

Yet, many things in nature do have approximate normal distributions, so normal distributions play a key role in statistics. Most of the classical statistical procedures assume that one has sampled from a population having an approximate distribution. In addition, it will be seen later than the CLT tells us in many of these cases that the quantities used for statistical estimation are approximately normal, even if the data they are calculated from are not.

Recall from above that the gamma distribution, or at least the Erlang, arises as a sum of independent random variables. Thus the Central Limit Theorem implies that the gamma distribution should be approximately normal for large (integer) values of r. We see in Figure 6.4 that even with $r = 10$ it is rather close to normal.

---

[6]The reader will notice that we have $\sigma$ in the above calculation, whereas the Chebychev analysis used the sample standard deviation $s$. The latter is only an estimate of the former, but one can show that the probability calculation is still valid.

## 9.10    The Chi-Squared Family of Distributions

### 9.10.1    Density and Properties

Let $Z_1, Z_2, ..., Z_k$ be independent $N(0, 1)$ random variables. Then the distribution of

$$Y = Z_1^2 + ... + Z_k^2 \qquad (9.23)$$

is called *chi-squared with k degrees of freedom*. We write such a distribution as $\chi_k^2$. Chi-squared is a one-parameter family of distributions, and arises quite frequently in classical statistical significance testing.

We can derive the mean of a chi-squared distribution as follows. First,

$$EY = E(Z_1^2 + ... + Z_k^2) = kE(Z_1^2) \qquad (9.24)$$

Well, $E(Z_1^2)$ sounds somewhat like variance, suggesting that we use (4.4). This works:

$$E(Z_1^2) = Var(Z_1) + [E(Z_1)]^2 = Var(Z_1) = 1 \qquad (9.25)$$

Then $EY$ in (9.23) is $k$. One can also show that $Var(Y) = 2k$.

It turns out that chi-squared is a special case of the gamma family in Section 6.7.4 below, with $r = k/2$ and $\lambda = 0.5$.

The R functions **dchisq()**, **pchisq()**, **qchisq()** and **rchisq()** give us the density, cdf, quantile function and random number generator for the chi-squared family. The second argument in each case is the number of degrees of freedom. The first argument is the argument to the corresponding math function in all cases but **rchisq()**, in which it is the number of random variates to be generated.

For instance, to get the value of $f_X(5.2)$ for a chi-squared random variable having 3 degrees of freedom, we make the following call:

```
> dchisq(5.2,3)
[1] 0.06756878
```

## 9.10.2 Example: Error in Pin Placement

Consider a machine that places a pin in the middle of a flat, disk-shaped object. The placement is subject to error. Let $X$ and $Y$ be the placement errors in the horizontal and vertical directions, respectively, and let $W$ denote the distance from the true center to the pin placement. Suppose $X$ and $Y$ are independent and have normal distributions with mean 0 and variance 0.04. Let's find $P(W > 0.6)$.

Since a distance is the square root of a sum of squares, this sounds like the chi-squared distribution might be relevant. So, let's first convert the problem to one involving squared distance:

$$P(W > 0.6) = P(W^2 > 0.36) \tag{9.26}$$

But $W^2 = X^2 + Y^2$, so

$$P(W > 0.6) = P(X^2 + Y^2 > 0.36) \tag{9.27}$$

This is not quite chi-squared, as that distribution involves the sum of squares of independent $N(0,1)$ random variables. But due to the normal family's closure under affine transformations (page 198), we know that $X/0.2$ and $Y/0.2$ do have $N(0,1)$ distributions. So write

$$P(W > 0.6) = P[(X/0.2)^2 + (Y/0.2)^2 > 0.36/0.2^2] \tag{9.28}$$

Now evaluate the right-hand side:

```
> 1 - pchisq(0.36/0.04,2)
[1] 0.01110900
```

## 9.10.3 Importance in Modeling

This distribution family does not come up directly in applications nearly so often as, say, the binomial or normal distribution family.

But the chi-squared family is used quite widely in statistical applications. As will be seen in our chapters on statistics, many statistical methods involve a sum of squared normal random variables.[7]

---

[7]The motivation for the term *degrees of freedom* will be explained in those chapters too.

### 9.10.4    Relation to Gamma Family

One can show that the chi-square distribution with $d$ degrees of freedom is a gamma distribution, with $r = d/2$ and $\lambda = 0.5$.

# 9.11    Mathematical Complements

### 9.11.1    Convergence in Distribution, and the Precisely-Stated CLT

The statement of Theorem 18 is not mathematically precise. We will fix it here.

**Definition 19** *A sequence of random variables $L_1, L_2, L_3, \ldots$ converges in distribution to a random variable $M$ if*

$$\lim_{n \to \infty} P(L_n \leq t) = P(M \leq t), \text{ for all } t \tag{9.29}$$

*In other words, the cdfs of the $L_i$ converge pointwise to that of $M$.*

The formal statement of the CLT is:

**Theorem 20** *Suppose $X_1, X_2, \ldots$ are independent random variables, all having the same distribution which has mean $m$ and variance $v^2$. Then*

$$Z = \frac{X_1 + \ldots + X_n - nm}{v\sqrt{n}} \tag{9.30}$$

*converges in distribution to a $N(0,1)$ random variable.*

Note by the way, that these random variables need not be defined on the same probability space. As noted, the hypothesis of the theorem merely says that the cdf of $L_n$ converges (pointwise) to that of $M$.

Similarly, the conclusion of the theorem does not say anything about densities. It does not state that the density of $L_n$ (even if it is a continuous random variable) converges to the $N(0,1)$ density, though there are various *local limit theorems* for that.

## 9.12 Computational Complements

### 9.12.1 Example: Generating Normal Random Numbers

How do normal random number generators such as **rnorm()** work? While in principle Section 6.9.2 could be used, the lack of a closed-form expression for $\Phi^{-1}()$ makes that approach infeasible. Instead, we can exploit the relation between the normal family and exponential distribution, as follows.

Let $Z_1$ and $Z_2$ be two independent $N(0,1)$ random variables, and define $W = Z_1^2 + Z_2^2$. By definition, $W$ has a chi-squared distribution with 2 degrees of freedom, and from Section 9.10, we know that $W$ has a gamma distribution with $r = 1$ and $\lambda = 0.5$.

In turn, we know from Section 6.7.4 that that distribution is actually just an exponential distribution with the same value of $\lambda$. This is quite fortuitous, since Section 6.9.2 *can* be used in this case; in fact, we saw in that section how to generate exponentially distributed random variables.

And there's more. Think of plotting the pair $(Z_1, Z_2)$ in the X-Y plane, and the angle $\theta$ formed by the point $(Z_1, Z_2)$ with the X axis:

$$\theta = \tan^{-1}\left(\frac{Z_2}{Z_1}\right) \tag{9.31}$$

Due to the symmetry of the situation, $\theta$ has a uniform distribution on $(0, 2\pi)$. Also,

$$Z_1 = \sqrt{W}\cos(\theta), \quad Z_2 = \sqrt{W}\sin(\theta) \tag{9.32}$$

Putting all this together, we can generate a pair of independent $N(0,1)$ random variates via the code

```
genn01 <- function() {
    theta <- runif(1,0,2*pi)
    w <- rexp(1,0.5)
    sw <- sqrt(w)
    c(sw*cos(theta),sw*sin(theta))
}
```

Note that we "get two for the price of one." If we need, say, 1000 random normal variates, we call the above function 500 times. Or better, use vectors:

```
genn01 <- function(n) {
   theta <- runif(n,0,2*pi)
   w <- rexp(n,0.5)
   sw <- sqrt(w)
   c(sw*cos(theta),sw*sin(theta))
}
```

This gives us $2n$ N(0,1) variates, so we call the function with $n = 250$ if we need 500 of them.

By the way, this method is called the *Box-Müller Transformation*. Here Box is the same rescercher in the quote at the start of Chapter 8.

## 9.13  Exercises

**Mathematical problems:**

**1.**  Continuing the Jill example in Section 9.5, suppose there is never an intrusion, i.e., all logins are from Jill herself. Say we've set our network intrusion monitor to notify us every time Jill logs in and accesses 535 or more disk sectors. In what proportion of all such notifications will Jill have accessed at least 545 sectors?

**2.**  Consider a certain river, and $L$, its level (in feet) relative to its average. There is a flood whenever $L > 8$, and it is reported that 2.5% of days have flooding. Let's assume that the level $L$ is normally distributed; the above information implies that the mean is 0. Suppose the standard deviation of $L$, $\sigma$, goes up by 10%. How much will the percentage of flooding days increase?

**3.**  Suppose the number of bugs per 1,000 lines of code has a Poisson distribution with mean 5.2. Find the approximate probability of having more than 106 bugs in 20 sections of code, each 1,000 lines long. Assume the different sections act independently in terms of bugs.

**4.**  Find $E(Z^4)$, where $Z$ has a $N(0,1)$ distribution. Hint: Use the material in Section 9.10.

**5.**  It can be shown that if the parent population has a normal distribution with variance $\sigma^2$, then the scaled sample variance $s^2/\sigma^2$ (standard version, with an $n - 1$ denominator), has a chi-squared distribution with $n - 1$ degrees of freedom.

 

   (a)  Find the MSE of $s^2$.

   (b)  Find the optimal denominator in terms of MSE. (It will not neces-
       sarily be either of the values we've discussed, $n$ and $n - 1$.)

**Computational and data problems:**

**6.**  Consider the setting of Section 9.6. Use simulation to find $E(N|N \geq 25)$.

**7.**  Use simulation to assess how accurate the CLT approximation was in Section 9.7.1.

**8.**  Suppose we model light bulb lifetimes as having a normal distribution with mean and standard deviation 500 and 50 hours, respectively. Give a loop-free R expression for finding the value of $d$ such that 30% of all bulbs have lifetime more than $d$.

# Chapter 10

# Introduction to Statistical Inference

*Statistical inference* involves careful (and limited) extrapolation from samples to populations. During an election, for instance, a survey organization might poll a sample of voters, then report, say, "Candidate Jones' support is 56.2%, with a margin of error of 3.5%." We'll discuss exactly what that means shortly, but for now the point is that the 3.5% figure recognizes that 0.562 is merely a sample estimate, not a population value, and it attempts to indicate how accurate the estimate might be. This chapter will cover such issues.

## 10.1 The Role of Normal Distributions

Classical statistics — going back 100 years or more — relies heavily on the assumption of normally distributed populations. Say for instance we are studying corporations' annual revenue $R$. The assumption is that $f_R$ would have the familiar bell shape. We know, though, that ths assumption cannot be precisely correct. If $R$ had a normal distribution, it would take on values from $-\infty$ to $\infty$. Revenue can't be negative (though profit can), and there is no corportion with revenue of, for example, $10^{50}$ dollars.

These methods are still in wide use today, but they work reasonably well on non-normal populations anyway. This is due to the Central Limit Theorem. No wonder they call it central!

Specifically, the key is what was presented in Secction 9.8. For convenience, let's repeat the main point here:

**Approximate distribution of (centered and scaled) $\overline{X}$:**

The quantity

$$Z = \frac{\overline{X} - \mu}{\sigma/\sqrt{n}} \qquad (10.1)$$

has an approximately $N(0,1)$ distribution, where $\sigma^2$ is the population variance.

So, even though a histogram of a sample of 1000 $R$ values may be very skewed, if we take, say 500 samples of size 1000 each, then plot a histogram of the resulting 500 sample means $\overline{R}$, the latter plot will look approximately bell-shaped!

So, let's get right to work, applying this.

## 10.2   Confidence Intervals for Means

We are now set to make use of the infrastructure that we've built up. Everything will hinge on understanding that the sample mean is a random variable, with a known approximate distribution (i.e., normal).

### 10.2.1   Basic Formulation

So, suppose we have a random sample $X_1, ..., X_n$ from some population with mean $\mu$ and variance $\sigma^2$ (but NOT necessarily normally distributed). Recall that (10.1) has an approximate $N(0,1)$ distribution. We will be interested in the central 95% of the distribution $N(0,1)$. Due to symmetry, that distribution has 2.5% of its area in the left tail and 2.5% in the right one. Through the R call **qnorm(-0.025)**, or by consulting a $N(0,1)$ cdf table in a book, we find that the cutoff points are at -1.96 and 1.96. In other words, if some random variable $T$ has a $N(0,1)$ distribution, then $P(-1.96 < T < 1.96) = 0.95$.

Thus

$$0.95 \approx P\left(-1.96 < \frac{\overline{X} - \mu}{\sigma/\sqrt{n}} < 1.96\right) \tag{10.2}$$

(Note the approximation sign.) Doing a bit of algebra on the inequalities yields

$$0.95 \approx P\left(\overline{X} - 1.96\frac{\sigma}{\sqrt{n}} < \mu < \overline{X} + 1.96\frac{\sigma}{\sqrt{n}}\right) \tag{10.3}$$

Now remember, not only do we not know $\mu$, we also don't know $\sigma$. But we can estimate it, as we saw, via (7.14). One can show[1] that (10.3) is still valid if we substitute $s$ for $\sigma$, i.e.,[2]

$$0.95 \approx P\left(\overline{X} - 1.96\frac{s}{\sqrt{n}} < \mu < \overline{X} + 1.96\frac{s}{\sqrt{n}}\right) \tag{10.4}$$

In other words, we are about 95% sure that the interval

$$(\overline{X} - 1.96\frac{s}{\sqrt{n}}, \overline{X} + 1.96\frac{s}{\sqrt{n}}) \tag{10.5}$$

contains $\mu$. This is called a 95% *confidence interval* for $\mu$.

Note the key connection to standard error, Section 7.7. Rephrasing the above in terms of standard error, we have:

An approximate 95% confidence interval for $\mu$ is $\overline{X}$ plus or minus 1.96 times the standard error of $\overline{X}$.

Even better, the same derivation above shows:

### Confidence Intervals from Approximately Normal Estimators

---

[1]This uses advanced probability theory, including Slutsky's Theorem, which states, roughly, that if $K_n$ converges in distribution to $K$ and the sequence of $L_n$ converges to a constant $c$, then $K_n/L_n$ converges in distribution to $K/c$. Here the $L_n$ will be $s$ and $c$ will be $\sigma$.

[2]Remember, all this is approximate. The approximation gets better as $n$ increases. But for any particular $n$ the approximation will be better with $\sigma$ than with $s$.

Suppose we are estimating some parameter $\theta$ in a parametric family, with an estimator $\widehat{\theta}$. If $\widehat{\theta}$ is approximately normally distributed,[3] then an approximate 95% confidence interval for $\theta$ is

$$\widehat{\theta} \pm 1.96 \text{ s.e.}(\widehat{\theta}) \tag{10.6}$$

Forming confidence intervals in terms of standard errors will come up repeatedly, both throughout this chapter and the ones that follow. This is because many estimators, not just $\overline{X}$, can be shown to be approximately normal. For example, most Maximum Likelihood Estimators have that property, so it ie easy to derive confidence intervals from them.

## 10.3   Example: Pima Diabetes Study

Let's look at the Pima data in Section 7.8, where we began a comparison of diabetic and nondiabetic women. Recall our notation: $\mu_1$ and $\sigma_1^2$ denote the population mean and variance of Body Mass Index **BMI** among diabetics, with $\mu_0$ and $\sigma_0^2$ being the corresponding quantities for the nondiabetics.

We are interested in estimating the difference $\theta = \mu_1 - \mu_0$. Our $\widehat{\theta}$ was $\overline{U} - \overline{V}$, which we found to be 35.14 - 30.30 = 4.84, with a standard error of 0.56. It definitely looked like the diabetics have higher BMI values, but on the other hand we know that those values are subject to sampling error. Is BMI substantially higher among diabetics in the population?

(Note the word *substantially*. We are not asking simply whether $\mu_1 > \mu_0$. If the difference were, say, 0.0000001, the two means would be essentially the same. This will be a key point later in this chapter.)

So, we form a confidence interval,

$$4.84 \pm 1.96(0.56) = (3.74, 5.94) \tag{10.7}$$

Now we have a range, an *interval estimate*, rather than just the *point estimate* 4.84. By presenting the results as an interval, we are recognizing that we are only dealing with sample estimates. We present a *margin of error* — the radius of the interval — 1.96(0.56) = 0.28.

---

[3]Technically this means that the approximate normality comes from the Central Limit Theorem.

The interval is rather wide, but it does indicate that the diabetics have a substantially higher BMI value, on average.

## 10.4 Example: Humidity Data

Consider the humidity data in Section 8.4.4. Recall that we fit a beta model there, estimating the parameters $\alpha$ and $\beta$ via MLE. Let's find a confidence interval for the former.

The value of $\widehat{\alpha}$ was 6.439, with a standard error of 0.334. So, our interval is

$$6.439 \pm 1.96(0.334) = (5.784, 7.094) \tag{10.8}$$

## 10.5 Meaning of Confidence Intervals

The key distinction between statistics and pure mathematics is that in the former, interpretation is of the utmost importance. Statistics professors tend to be quite fussy about this. What does a confidence interval really mean?

### 10.5.1 A Weight Survey in Davis

Consider the question of estimating the mean weight, denoted by $\mu$, of all adults in the city of Davis. Say we sample 1000 people at random, and record their weights, with $W_i$ being the weight of the $i^{th}$ person in our sample.

**Now remember, we don't know the true value of that population mean, $\mu$ — again, that's why we are collecting the sample data, to estimate $\mu$! Our estimate will be our sample mean, $\overline{W}$. But we don't know how accurate that estimate might be. That's the reason we form the confidence interval, as a gauge of the accuracy of $\overline{W}$ as an estimate of $\mu$.**

Say our interval (10.5) turns out to be (142.6,158.8). We say that we are about 95% confident that the mean weight $\mu$ of all adults in Davis is contained in this interval. **What does this mean?**

Say we were to perform this experiment many, many times, recording the results in a notebook: We'd sample 1000 people at random, then record our interval $(\overline{W} - 1.96\frac{s}{\sqrt{n}}, \overline{W} + 1.96\frac{s}{\sqrt{n}})$ on the first line of the notebook. Then we'd sample another 1000 people at random, and record what interval we got that time on the second line of the notebook. This would be a different set of 1000 people (though possibly with some overlap), so we would get a different value of $\overline{W}$ and so, thus a different interval; it would have a different center and a different radius. Then we'd do this a third time, a fourth, a fifth and so on.

Again, each line of the notebook would contain the information for a different random sample of 1000 people. There would be two columns for the interval, one each for the lower and upper bounds. And though it's not immediately important here, note that there would also be columns for $W_1$ through $W_{1000}$, the weights of our 1000 people, and columns for $\overline{W}$ and s.

Now here is the point: Approximately 95% of all those intervals would contain $\mu$, the mean weight in the entire adult population of Davis. The value of $\mu$ would be unknown to us — once again, that's why we'd be sampling 1000 people in the first place — but it does exist, and it would be contained in approximately 95% of the intervals. **This latter point is what we mean when we say we are 95% "sure" that $\mu$ is contained in the particular interval we form.**

As a variation on the notebook idea, think of what would happen if you and 99 friends each do this experiment. Each of you would sample 1000 people and form a confidence interval. Since each of you would get a different sample of people, you would each get a different confidence interval. What we mean when we say the confidence level is 95% is that of the 100 intervals formed — by you and 99 friends — about 95 of them will contain the true population mean weight. Of course, you hope you yourself will be one of the 95 lucky ones! But remember, you'll never know whose intervals are correct and whose aren't.

**Now remember, in practice we only take *one* sample of 1000 people. Our notebook idea here is merely for the purpose of understanding what we mean when we say that we are about 95% confident that one interval we form does contain the true value of $\mu$.**

There is more on the interpretation of confidence intervals in Section 10.17.1.

## 10.6  Confidence Intervals for Proportions

So we know how to find confidence intervals for means. How about proportions?

For example, in an election opinion poll, we might be interested in the proportion $p$ of people in the entire population who plan to vote for candidate A. We take as our estimate $\hat{p}$, the corresponding proportion in our sample. How can we form a confidence interval?

Well, remember, a proportion *is* a mean. We found that in (4.36). Or, on the sample level, the $X_i$ here are 1s and 0s (1 if favor of candidate A, 0 if not). Now, what is the average of a bunch of 0s and 1s? The numerator in the sample mean $\overline{X}$ will be the sum of the $X_i$, which will simply be the count of the number of 1s. Dividing by the sample size, we get the sample proportion. In other words,

$$\hat{p} = \overline{X} \tag{10.9}$$

Moreover, consider $s^2$. Look at (7.5.2). Since the $X_i$ are all 0s and 1s, then $X_i^2 = X_i$. That means (7.5.2) is

$$s^2 = \hat{p} - \hat{p}^2 = \hat{p}(1 - \hat{p}) \tag{10.10}$$

So, the standard error of $\hat{p}$ is

$$\frac{s}{\sqrt{n}} = \sqrt{\frac{\hat{p}(1 - \hat{p})}{n}} \tag{10.11}$$

We then use (10.5) to obtain our interval,

$$\hat{p} + 1.96 \sqrt{\frac{\hat{p}(1 - \hat{p})}{n}} \tag{10.12}$$

In the case of the difference between two proportions $\hat{p}_1 - \hat{p}_2$, the standard error, combining the above and (7.22), is

$$\sqrt{\frac{\hat{p}_1(1 - \hat{p}_1)}{n_1} + \frac{\hat{p}_2(1 - \hat{p}_2)}{n_2}} \tag{10.13}$$

## 10.6.1    Example: Machine Classification of Forest Covers

*Remote sensing* is machine classification of type from variables observed aerially, typically by satellite. The application we'll consider here involves forest cover type for a given location; there are seven different types [4]. The dataset is in the UC Irvine Machine Learning Repository [12].

Direct observation of the cover type is either too expensive or may suffer from land access permission issues. So, we wish to guess cover type from other variables that we can more easily obtain.

One of the variables was the amount of hillside shade at noon, which we'll call HS12. *Here's our goal:* Let $\mu_1$ and $\mu_2$ be the population mean HS12 among sites having cover types 1 and 2, respectively. If $\mu_1 - \mu_2$ is large, then HS12 would be a good predictor of whether the cover type is 1 or 2.

So, we wish to estimate $\mu_1 - \mu_2$ from our data, in which we do know cover type. There were over 580,000 observations. We'll use R's **tapply()** function (Section 7.12.1.3):

```
> tapply(cvr[,8],cvr[,55],mean)
        1         2         3         4         5
 223.4302  225.3266  215.8265  216.9971  219.0358
        6         7
 209.8277  221.7460
```

So, $\widehat{\mu}_1 = 223.43$ and $\widehat{\mu}_2 = 225.33$. We'll need the values of $s^2$ and sample sizes as well:

```
> tapply(cvr[,8],cvr[,55],var)
        1         2         3         4         5
 329.7829  342.6033  778.7232  437.5353  620.6744
        6         7
 596.2166  400.0025
> tapply(cvr[,8],cvr[,55],length)
      1       2       3       4       5       6       7
 211840  283301   35754    2747    9493   17367   20510
```

As in (7.22), the standard error of $\widehat{\mu}_1 - \widehat{\mu}_2$ is then

$$\sqrt{\frac{329.78}{211840} + \frac{342.60}{283301}} = 0.05 \tag{10.14}$$

So our confidence interval for the difference between the mean HS12 values in the two populaitons is

$$223.43 - 225.33 \pm 1.96(0.05) = (-2.00, -1.80) \qquad (10.15)$$

Given that HS12 values are in the 200 range (see the sample means), we see from the confidence interval that the difference between the two population means is minuscule. It does not appear that HS12, at least by itself, will help us predict whether the cover type is 1 vs. 2.

This is a great illustration of an important principle, discussed in detail in Section 10.15.

As another illustration of confidence intervals, let's find one for the difference in population proportions of sites that have cover types 1 and 2. To obtain our sample estimates, we run

```
> ni <- tapply(cvr[,8],cvr[,55],length)
> ni/sum(ni)
          1              2              3              4
0.364605206  0.487599223  0.061537455  0.004727957
          5              6              7
0.016338733  0.029890949  0.035300476
```

So,

$$\widehat{p}_1 - \widehat{p}_2 = 0.365 - 0.488 = -0.123 \qquad (10.16)$$

The standard error of this quantity, from (10.13), is

$$\sqrt{0.365 \cdot (1 - 0.365)/211840 + 0.488 \cdot (1 - 0.488)/283301} = 0.001 \qquad (10.17)$$

That gives us a confidence interval of

$$-0.123 \pm 1.96 \cdot 0.001 = (-0.121, -0.125) \qquad (10.18)$$

Needless to say, with this large sample size, our estimate is likely quite accurate.

Assuming the data are a random sample for the forest population of interest, there appear to be substantially more sites of type 2.

# 10.7   The Student-t Distribution

*Far better an approximate answer to the right question, which is often vague, than an exact answer to the wrong question, which can always be made precise*—John Tukey, pioneering statistician at Bell Labs and Princeton University

Classicly, analysts have used the *Student t-distribution* for inference. That is the name of the distribution of the quantity

$$ T = \frac{\overline{X} - \mu}{\tilde{s}/\sqrt{n-1}} \tag{10.19} $$

where $\tilde{s}^2$ is the version of the sample variance in which we divide by $n-1$ instead of by $n$.

Note carefully that we are assuming that the $X_i$ themselves—not just $\overline{X}$—have a normal distribution. In other words, if we are studying human weight, say, then the assumption is that weight follows an exact bell-shaped curve. The exact distribution of $T$ is called the *Student t-distribution with n-1 degrees of freedom*. These distributions thus form a one-parameter family, with the degrees of freedom being the parameter.

The general definition of the Student-t family is the distribution of ratios $U/\sqrt{V/k}$, where

- $U$ has a $N(0,1)$ distribution

- $V$ has a chi-squared distribution with $k$ degrees of freedom

- U and V are independent

It can be shown that in (10.19), if the sampled population has a normal distribution, then $(\overline{X} - \mu)/\sigma$ and $\tilde{s}^2/\sigma^2$ actually do satisfy the above conditions on U and V, respectively, with $k = n - 1$. (If we are forming a confidence interval for the difference of two means, the calculation of degrees of freedom becomes more complicated, but it is not important here.)

This distribution has been tabulated. In R, for instance, the functions **dt()**, **pt()** and so on play the same roles as **dnorm()**, **pnorm()** etc. do for the normal family. The call **qt(0.975,9)** returns 2.26. This enables us to get a confidence interval for $\mu$ from a sample of size 10, at EXACTLY a 95% confidence level, rather than being at an APPROXIMATE 95% level as we have had here, as follows.

We start with (10.2), replacing 1.96 by 2.26, $(\bar{X} - \mu)/(\sigma/\sqrt{n})$ by T, and $\approx$ by $=$. Doing the same algebra, we find the following confidence interval for $\mu$:

$$(\bar{X} - 2.26\frac{\tilde{s}}{\sqrt{10}}, \bar{X} + 2.26\frac{\tilde{s}}{\sqrt{10}}) \qquad (10.20)$$

Of course, for general $n$, replace 2.26 by $t_{0.975,n-1}$, the 0.975 quantile of the t-distribution with n-1 degrees of freedom.

We do not use the t-distribution in this book, because:

- It depends on the parent population having an exact normal distribution, which is never really true. In the Davis case, for instance, people's weights are approximately normally distributed, but definitely not exactly so. For that to be exactly the case, some people would have to have weights of say, a billion pounds, or negative weights, since any normal distribution takes on all values from $-\infty$ to $\infty$.

- For large $n$, the difference between the t-distribution and $N(0,1)$ is negligible anyway. That wasn't true in the case $n = 10$ above, where our confidence interval multiplied the standard error by 2.26 instead of 1.96 as we'd seen earlier. But for $n = 50$, the 2.26 already shrinks to 2.01, and for $n = 100$, it is 1.98.

In other words, for small $n$, the claim of exactness for t-based inference is usually unwarranted, and for large $n$, the difference between the t-distribution and N(0,1) is small anyway. So we might as well just use the latter, as we have been doing in this chapter.

## 10.8 Introduction to Significance Tests

On the one hand, the class of methods known as *significance tests* form the very core of statistics. Open any journal in science, medicine, psychology, economics and so on, and you will find significance tests in virtually every article.

On the other hand, in 2016 the American Statistical Association issued its first-ever policy statement [41], asserting that significance tests are widely overused and misinterpreted. It noted, though:

Let us be clear. Nothing in the ASA statement is new. Statisticians and others have been sounding the alarm about these matters for decades, to little avail.

Then in 2019, an article was published in *Nature*, one of the two most prestigious science journals in the world, echoing the ASA statement and taking it one step further [1].

Well, then, what do significance tests, this core statistical methodology, actually do, and why were this very august scientific body and equally-august scientific journal "sounding the alarm"?

To answer, let's look at a simple example of deciding whether a coin is fair, i.e., has heads probability 0.5.

## 10.9    The Proverbial Fair Coin

Suppose (just for fun, but with the same pattern as in more serious examples) you have a coin that will be flipped at the Super Bowl to see who gets the first kickoff.[4] You want to assess for "fairness." Let $p$ be the probability of heads for this coin. A fair coin would have $p = 0.5$.

You could toss the coin, say, 100 times, and then form a confidence interval for $p$. The width of the interval would tell you the margin of error, i.e., it tells you whether 100 tosses were enough for the accuracy you want, and the location of the interval would tell you whether the coin is "fair" enough.

For instance, if your interval were (0.49,0.54), you might feel satisfied that this coin is reasonably fair. In fact, **note carefully that even if the interval were, say, (0.502,0.506), you would still consider the coin to be reasonably fair**; the fact that the interval did not contain 0.5 is irrelevant, as the entire interval would be reasonably near 0.5.

However, this process would not be the way it's traditionally done. Most users of statistics would use the toss data to test the **null hypothesis**

$$H_0 : p = 0.5 \qquad (10.21)$$

---

[4]We'll assume slightly different rules here. The coin is not "called." Instead, it is agreed beforehand that if the coin comes up heads, Team A will get the kickoff, and otherwise it will be Team B.

against the **alternate hypothesis**

$$H_A : p \neq 0.5 \tag{10.22}$$

For reasons that will be explained below, this procedure is called **significance testing**.

## 10.10   The Basics

Here's how significance testing works.

The approach is to consider $H_0$ "innocent until proven guilty," meaning that we assume $H_0$ is true unless the data give strong evidence to the contrary.

The basic plan of attack is this:

> We will toss the coin $n$ times. Then we will believe that the coin is fair unless the number of heads is "suspiciously" extreme, i.e., much less than $n/2$ or much more than $n/2$.

As before, let $\widehat{p}$ denote the sample proportion, in this case the proportion of heads in our sample of $n$ tosses, and recall that

$$\frac{\widehat{p} - p}{\sqrt{\frac{1}{n} \cdot p(1-p)}} \tag{10.23}$$

has an approximate $N(0,1)$ distribution.

But remember, we are going to assume $H_0$ for now, until and unless we find strong evidence to the contrary. So, let's substitute $p = 0.5$ in (10.23), yielding that

$$Z = \frac{\widehat{p} - 0.5}{\sqrt{\frac{1}{n} \cdot 0.5(1-0.5)}} \tag{10.24}$$

has an approximate $N(0,1)$ distribution (again, under the assumption that $H_0$ is true).

Now recall from the derivation of (10.5) that -1.96 and 1.96 are the lower- and upper-2.5% points of the $N(0,1)$ distribution. Thus, again under $H_0$,

$$P(Z < -1.96 \text{ or } Z > 1.96) \approx 0.05 \qquad (10.25)$$

Now here is the point: After we collect our data, in this case by tossing the coin $n$ times, we compute $\hat{p}$ from that data, and then compute $Z$ from (10.24). If $Z$ is smaller than -1.96 or larger than 1.96, we reason as follows:

> Hmmm. If $H_0$ were true, $Z$ would stray that far from 0 only 5% of the time. So, either I have to believe that a rare event has occurred, or I must abandon my assumption that $H_0$ is true. I choose to abandon the assumption.

For instance, say $n = 100$ and we get 62 heads in our sample. That gives us $Z = 2.4$, in that "rare" range. We then *reject* $H_0$, and announce to the world that this is an unfair coin. We say, "The value of $p$ is significantly different from 0.5."

Just as analysts commonly take 95% for their confidence intervals, it is standard to use 5% as our "suspicion criterion"; this is called the *significance level*, typically denoted $\alpha$. One common statement is "We rejected $H_0$ at the 5% level."

The word *significant* is misleading. It should NOT be confused with *important*. It simply is saying we don't believe the observed value of Z, 2.4, is a rare event, which it would be under $H_0$; we have instead decided to abandon our belief that $H_0$ is true.

On the other hand, suppose we get 47 heads in our sample. Then $Z = -0.60$. Again, taking 5% as our significance level, this value of $Z$ would not be deemed suspicious, as it is in a range of values, (-1.96,1.96), that would occur frequently under $H_0$. We would then say "We accept $H_0$ at the 5% level," or "We find that $p$ is not significantly different from 0.5."

Note by the way that $Z$ values of -1.96 and 1.96 correspond getting $50 - 1.96 \cdot 0.5 \cdot \sqrt{100}$ or $50 + 1.96 \cdot 0.5 \cdot \sqrt{100}$ heads, i.e., roughly 40 or 60. In other words, we can describe our rejection rule to be "Reject if we get fewer than 40 or more than 60 heads, out of our 100 tosses."

## 10.11    General Normal Testing

At the end of Section 10.2.1, we developed a method of constructing confidence intervals for general approximately normally distributed estimators. Now we do the same for significance testing.

Suppose $\widehat{\theta}$ is an approximately normally distributed estimator of some population value $\theta$. Then to test $H_0 : \theta = c$, form the test statistic

$$Z = \frac{\widehat{\theta} - c}{s.e.(\widehat{\theta})} \tag{10.26}$$

where $s.e.(\widehat{\theta})$ is the standard error of $\widehat{\theta}$, and proceed as before:

> Reject $H_0 : \theta = c$ at the significance level of $\alpha = 0.05$ if $|Z| \geq 1.96$.

## 10.12    The Notion of "p-Values"

Recall the coin example in Section 10.8, in which we got 62 heads, i.e., $Z = 2.4$. Since 2.4 is considerably larger than 1.96, our cutoff for rejection, we might say that in some sense we not only rejected $H_0$, we actually <u>strongly</u> rejected it.

To quantify that notion, we compute something called the *observed significance level*, more often called the *p-value*.

We ask,

> We rejected $H_0$ at the 5% level. Clearly, we would have rejected it even at some smaller — thus more stringent — levels. What is the smallest such level? Call this the p-value of the test.

By checking a table of the $N(0, 1)$ distribution, or by calling **pnorm(2.40)** in R, we would find that the $N(0, 1)$ distribution has area 0.008 to the right of 2.40, and of course by symmetry there is an equal area to the left of -2.40. That's a total area of 0.016. In other words, we would have been able to reject $H_0$ even at the much more stringent significance level of 0.016 (the 1.6% level) instead of 0.05. So, $Z = 2.40$ would be considered even more "significant" than $Z = 1.96$. In the research community it is customary to say, "The p-value was 0.016."[5] The smaller the p-value, the more "significant" the results are considered.

In computer output or research reports, we often see small p-values being denoted by asterisks. There is generally one asterisk for p under 0.05, two for p less than 0.01, three for 0.001, etc. The more asterisks, the more significant the data is supposed to be. See for instance the R regression output on page 316.

## 10.13    What's Random and What Is Not

It is crucial to keep in mind that $H_0$ is not an event or any other kind of random entity. The coin in our example either has $p = 0.5$ or it doesn't. If we repeat the experiment, we will get a different value of $X$, the number of heads out of 100, but $p$ doesn't change; it's still the same coin! So for example, it would be wrong and meaningless to speak of the "probability that $H_0$ is true."

## 10.14    Example: The Forest Cover Data

Let's test the hypothesis of equal population means,

$$H_0 : \mu_1 = \mu_2 \qquad\qquad (10.27)$$

in the Forest Cover data, Section 10.6.1.

In the context of Section 10.11, $\theta = \mu_1 - \mu_2$, $\widehat{\theta} = \widehat{\mu}_1 - \widehat{\mu}_2$, and $c = 0$.

In Section 10.6.1, the sample size was about 580,000.  As you will see below. significance testing is essentially useless for such large samples, so for illustration purposes let's see what may have occurred with a smaller sample size. We'll extract a subsample of size 1000, and pretend that that was our actual sample.

```
> cvr1000 <- cvr[sample(1:nrow(cvr),1000),]
> muhats <- tapply(cvr1000[,8],cvr1000[,55],mean)
```

---

[5]The 'p' in "p-value" of course stands for "probability," meaning the probably that a $N(0,1)$ random variable would stray as far, or further, from 0 as our observed $Z$ here. By the way, be careful not to confuse this with the quantity $p$ in our coin example, the probability of heads.

```
> muhats
        1         2         3         4         5
222.6823  225.5040  216.2264  205.5000  213.3684
        6         7
208.9524  226.7838
> diff <- muhats[1] - muhats[2]
> diff
        1
-2.821648
> vars <- tapply(cvr1000[,8],cvr1000[,55],var)
> ns <- tapply(cvr1000[,8],cvr1000[,55],length)
> se <- sqrt(vars[1]/ns[1] + vars[2]/ns[2])
> se
        1
1.332658
> z <- diff / se
> z
        1
-2.117309
> 2 * pnorm(z)
        1
0.03423363
```

So, we would reject the hypothesis of equal population means, with a p-value of about 0.03. Now let's see what happens with the larger sample size:

We had earlier found a confidence interval for $\mu_1 - \mu_2$,

$$223.43 - 225.33 \pm 1.96(0.05) = (-2.00, -1.80) \qquad (10.28)$$

The 0.05 value was the standard error of $\widehat{\mu}_1 - \widehat{\mu}_2$. Let's try a significance test, for the null hypothesis So,

$$Z = \frac{(223.43 - 225.33) - 0}{0.05} = -38.0 \qquad (10.29)$$

This number is "way off the chart," so the area to its left is infinitesimal. In the common parlance, the difference in HS12 values would be said to be "very highly significant," to say the least. Or more exhuberantly, "extremely fantastically significant." A researcher preparing a paper for an academic journal would jump for joy.

And yet...looking at the confidence interval above, we see that the difference in HS12 between cover types 1 and 2 is tiny when compared to the general size of HS12, in the 200s. Thus HS12 is not going to help us guess which cover type exists at a given location. In this sense, the difference is not "significant" at all. And this is why the American Statistical Association released their historic position paper, warning that p-values were overused and often misinterpreted.

## 10.15　Problems with Significance Testing

*Sir Ronald [Fisher] has befuddled us, mesmerized us, and led us down the primrose path* — Paul Meehl, professor of psychology and the philosophy of science, referring to Fisher, one of the major founders of statistical methodology

**Significance testing is a time-honored approach, used by tens of thousands of people every day.** But although significance testing is mathematically correct, many consider it to be at best noninformative and at worst seriously misleading.

### 10.15.1　History of Significance Testing

When the concept of significance testing, especially the 5% value for $\alpha$, was developed in the 1920s by Sir Ronald Fisher, many prominent statisticians opposed the idea — for good reason, as we'll see below. But Fisher was so influential that he prevailed, and thus significance testing became the core operation of statistics.

So, significance testing became entrenched in the field, in spite of being widely recognized as potentially problematic to this day. Most modern statisticians understand this, even if many continue to engage in the practice.[6] For instance, there is an entire chapter devoted to this issue in one of the best-selling elementary statistics textbooks in the US [17].

One of the authors of that book, Professor David Freedman of UC Berkeley, was commissioned to write a guide to statistics for judges [24]. The discussion there of potential problems with significance testing is similar to

---

[6]Many are forced to do so, e.g., to comply with government standards in pharmaceutical testing. My own approach in such situations is to quote the test results but then point out the problems, and present confidence intervals as well.

that of our next section here. These views are held by most statisticians, and led to the ASA statement cited above.

## 10.15.2   The Basic Issues

To begin with, it's questionable to test $H_0$ in the first place, because we almost always know *a priori* that $H_0$ is false.

Consider the coin example, for instance. No coin is absolutely perfectly balanced — e.g., the *bas relief* bust of Abraham Lincoln on the "heads" side of the US penny would seem to make that side heavier — and yet that is the question that significance testing is asking:

$$H_0 : p = 0.50000000000000000000000000000... \qquad (10.30)$$

We know before even collecting any data that the hypothesis we are testing is false, and thus it's nonsense to test it.

But much worse is this word "significant." Say our coin actually has $p = 0.502$. From anyone's point of view, that's a fair coin! But look what happens in (10.24) as the sample size $n$ grows. If we have a large enough sample, eventually the denominator in (10.24) will be small enough, and $\hat{p}$ will be close enough to 0.502, that $Z$ will be larger than 1.96 and we will declare that $p$ is "significantly" different from 0.5. But it isn't! Yes, 0.502 is different from 0.5, but NOT in any significant sense in terms of our deciding whether to use this coin in the Super Bowl.

The same is true for government testing of new pharmaceuticals. We might be comparing a new drug to an old drug. Suppose the new drug works only, say, 0.4% (i.e., 0.004) better than the old one. Do we want to say that the new one is "signficantly" better? This wouldn't be right, especially if the new drug has much worse side effects and costs a lot more (a given, for a new drug).

Note that in our analysis above, in which we considered what would happen in (10.24) as the sample size increases, we found that eventually *everything* becomes "signficiant"—even if there is no practical difference. This is especially a problem in computer applications of statistics, because they often use very large data sets.

That is what we saw in the forest cover example above. The p-value was essentially 0, yet the difference in population means was so small that it was negligible in terms of our goal of predicting cover type.

In all of these examples, the standard use of significance testing can result in our pouncing on very small differences that are quite insignificant to us, yet will be declared "significant" by the test.

Conversely, if our sample is too small, we can miss a difference that actually *is* significant — i.e., important to us — and we would declare that $p$ is NOT significantly different from 0.5. In the example of the new drug, this would mean that it would be declared as "not significantly better" than the old drug, even if the new one is much better but our sample size wasn't large enough to show it.

In summary, the basic problems with significance testing are

- $H_0$ is improperly specified. What we are really interested in here is whether $p$ is *near* 0.5, not whether it is *exactly* 0.5 (which we know is not the case anyway).

- Use of the word *significant* is grossly improper (or, if you wish, grossly misinterpreted).

### 10.15.3   Alternative Approach

*I was in search of a one-armed economist, so that the guy could never make a statement and then say: "on the other hand"* — President Harry S Truman

*If all economists were laid end to end, they would not reach a conclusion—* Irish writer George Bernard Shaw

Note carefully that this is not to say that we should not make a decision. We *do* have to decide, e.g., decide whether a new hypertension drug is safe or in this case decide whether this coin is "fair" enough for practical purposes, say for determining which team gets the kickoff in the Super Bowl. But it should be an informed decision.

In fact, the real problem with significance tests is that they **take the decision out of our hands**. They make our decision mechanically for us, not allowing us to interject issues of importance to us, such possible side effects in the drug case.

Forming a confidence interval is the more informative approach. In the coin example, for instance:

- The width of the interval shows us whether $n$ is large enough for $\widehat{p}$ to be reasonably accurate.

- The location of the interval tells us whether the coin is fair enough for our purposes.

Note that in making such a decision, we do NOT simply check whether 0.5 is in the interval. That would make the confidence interval reduce to a significance test, which is what we are trying to avoid. If for example the interval is (0.502,0.505), we would probably be quite satisfied that the coin is fair enough for our purposes, even though 0.5 is not in the interval.

On the other hand, say the interval comparing the new drug to the old one is quite wide and more or less equal positive and negative territory. Then the interval is telling us that the sample size just isn't large enough to say much at all.

In the movies, you see stories of murder trials in which the accused must be "proven guilty beyond the shadow of a doubt." But in most noncriminal trials, the standard of proof is considerably lighter, *preponderance of evidence*. This is the standard you must use when making decisions based on statistical data. Such data cannot "prove" anything in a mathematical sense. Instead, it should be taken merely as evidence. The width of the confidence interval tells us the likely accuracy of that evidence. We must then weigh that evidence against other information we have about the subject being studied, and then ultimately make a decision on the basis of the preponderance of all the evidence.

Yes, juries must make a decision. But they don't base their verdict on some formula. Similarly, you the data analyst should not base your decision on the blind application of a method that is usually of little relevance to the problem at hand—significance testing.

## 10.16   The Problem of "P-hacking"

The (rather recent) term *p-hacking* refers to the following abuse of statistics.[7]

---

[7]The term *abuse* here will not necessarily connote intent. It may occur out of ignorance of the problem.

## 10.16.1 A Thought Experiment

Say we have 250 pennies, and we wish to determine whether any are unbalanced, i.e., have probability $p$ of heads different from 0.5. (As noted earlier, we know *a priori* that none will have $p$ exactly equal to 0.5, but for the purpose of this thought experiment, let's put that aside for now.)

We investigate by tossing each coin 100 times, and testing the hypothesis $H_0 : p = 0.5$ for each coin, where $p$ is the probability of heads for that coin. Following the analysis in Section 10.8, if we get fewer than 40 heads or more than 60, we decide that coin is unbalanced. **The problem is that, even if all the coins are perfectly balanced, we eventually will have one that yields fewer than 40 or greater than 60 heads, just by accident.** We will then falsely declare this coin to be unbalanced.

For any particular penny, we have only only a 5% chance of falsely rejecting $H_0$, but *collectively* we have a problem: The probability that we have at least one false rejection among the 250 pennies is $1 - 0.95^{250} = 0.9999973$. So it is almost certain that we have at least one wrong conclusion.

Or, to give another frivolous example that still will make the point, say we are investigating whether there is any genetic component to a person's sense of humor. Is there a Humor gene? There are many, many genes to consider. Testing each one for relation to sense of humor is like checking each penny for being unbalanced: Even if there is no Humor gene, then eventually, just by accident, we'll stumble upon one that seems to be related to humor.

Of course the problem is the same for confidence intervals. If we compute a confidence interval for each of the 250 pennies, the chances are high that at least one of the intervals is seriously misleading.

There is no way to avoid the problem. The most important thing is to recognize that there is in fact this problem, and to realize, say, that if someone announces they've discovered a Humor gene, based on testing thousands and genes, the finding may be spurious.

## 10.16.2 Multiple Inference Methods

There are techniques called *multiple inference, simultaneous inference* or *multiple comparison* methods, to deal with p-hacking in performing statistical inference. See for example [21]. The simplest, *Bonferroni's Method*, just uses the following simple approach. If we wish to compute five confidence intervals with an *overall* confidence level of 95%, we must set the confidence level of each individal interval to 99%. This causes it to be wider, but safer.

Here is why Bonferroni works:

**Proposition 21** *Consider events* $A_i$, $i = 1, 2, ..., , k$. *Then*

$$P(A_1 \text{ or } A_2 \text{ or } ... \text{ or } A_k) \leq \sum_{i=1}^{k} P(A_i) \qquad (10.31)$$

In our example of setting five confidence intervals at the 99% level in order to achieve an overall level of at least 95%, we have $k = 5$, with $A_i$ being the event that the $i^{th}$ interval fails to contain the desired population quantity. In (10.31), the left-hand side is the probability that at least one CI fails, and the right-hand side is $5 \times 0.01 = 0.05$. So, we are bounding our failure rate by 0.05, as desired.

(10.31) is easily derived. It is immediately true for $k = 2$, by (1.5), since $P(A \text{ and } B) \geq 0$. Then use mathemtical induction.

# 10.17 Philosophy of Statistics

## 10.17.1 More about Interpretation of CIs

Some statistics instructors give students the odd warning, "You can't say that the probability is 95% that $\mu$ is IN the interval; you can only say that the probability is 95% confident that the interval CONTAINS $\mu$." This of course does not make sense; the following two statements are equivalent:

- "$\mu$ is in the interval"

- "the interval contains $\mu$"

Where did this seemingly unreasonable distinction between *in* and *contains* come from? Well, way back in the early days of statistics, some instructor was afraid that a statement like "The probability is 95% that $\mu$ is in the interval" would make it sound like $\mu$ is a random variable. Granted, that was a legitimate fear, because $\mu$ is not a random variable, and without proper warning, some learners of statistics might think incorrectly. The random entity is the interval (both its center and radius), not $\mu$; $\overline{X}$ and $s$

in (10.5) vary from sample to sample, so the interval is indeed the random object here, not $\mu$.

So, it was reasonable for teachers to warn students not to think $\mu$ is a random variable. But later on, some misguided instructor must have then decided that it is incorrect to say "$\mu$ is in the interval," and others then followed suit. They continue to this day, sadly.

A variant on that silliness involves saying that one can't say "The probability is 95% that $\mu$ is in the interval," because $\mu$ is either in the interval or not, so that "probability" is either 1 or 0! That is equally mushy thinking.

Suppose, for example, that I go into the next room and toss a coin, letting it land on the floor. I return to you, and tell you the coin is lying on the floor in the next room. I know the outcome but you don't. What is the probability that the coin came up heads? To me that is 1 or 0, yes, but to you it is 50%, in any practical sense.

It is also true in the "notebook" sense. If I do this experiment many times—go to the next room, toss the coin, come back to you, go to the next room, toss the coin, come back to you, etc., one line of the notebook per toss—then in the long run 50% of the lines of the notebook have Heads in the Outcome column.

The same is true for confidence intervals. Say we conduct many, many samplings, one per line of the notebook, with a column labeled Interval Contains Mu. Unfortunately, we ourselves don't get to see that column, but it exists, and in the long run 95% of the entries in the column will be Yes.

Finally, there are those who make a distinction between saying "There is a 95% probability that..." and "We are 95% confident that..." That's silly too. What else could "95% confident" mean if not 95% probability?

Consider the experiment of tossing two fair dice. The probability is 34/36, or about 94%, that we get a total that is different from 2 or 12. As we toss the dice, what possible distinction could be made between saying, "The probability is 94% that we will get a total between 3 and 11" and saying, "We are 94% confident that we will get a total between 3 and 11"? The notebook interpretation supports both phrasings, really. The words *probability* and *confident* should not be given much weight here; remember the quote at the beginning of our Chapter 1:

*I learned very early the difference between knowing the name of something and knowing something*—Richard Feynman, Nobel laureate in physics

### 10.17.1.1 The Bayesian View of Confidence Intervals

Recall the Bayesian philosophy, introduced in Section 8.7. Is there something like a confidence interval in that world? The answer is different yes, but it takes a different form, and of course has a different name.

In place of a CI, a Bayesian may compute the central 95% range of the posterior distribution, which is then termed a *credible interval.*

## 10.18  Exercises

**Mathematical problems:**

**1**. Suppose in (10.5) we use 1.80 instead of 1.96. What will the approximate confidence level be?

**2**. Say we are estimating a density, as in Chapter 8. Show how to form a confidence interval for the height of the density at the center of a bin. Apply your formula to the BMI data in that chapter.

**3**. It can be shown that if the parent population has a normal distribution with variance $\sigma^2$, then the scaled sample variance $s^2/\sigma^2$ (standard version, with an $n-1$ denominator, has a chi-squared distribution with $n-1$ degrees of freedom. Use this fact to derive an exact 95% confidence interval for $\sigma^2$. For convenience, make it a *one-sided* interval, say an upper bound, i.e., one from which we say "We are 95% confident that $\sigma^2 \leq c$."

**Computational and data problems:**

**4**. In the Forest Cover data, Section 10.14, find approximate confidence intervals for all seven mean HS12 values, with overall level of 95%. Use the Bonferroni method.

**5**. In the Forest Cover data, consider classes 1 and 2. Find an approximate 95% confidence interval for the difference in proportions of values over 240.

**6**. Load R's built-in dataset, **UCBAdmissions**, which arose in a graduate school admissions controversy at UC Berkeley. (See Section 14.5.1 on working with R's **"table"** class.) Plaintiffs at UC Berkeley contended that the university had been discriminating against female applicants. Find an approximate 95% confidence interval for the difference in male and female population admissions rates. Then find Bonferroni intervals for the male-female difference within each of the six academic departments. The conditional and unconditional results are at odds with each other, a famous example of *Simpson's Paradox*. Comment.

**7.** In the Bodyfat example, Section 8.9.2.2, find an approximate 95% confidence interval for $\beta$.

**8.** Suppose we take a random sample of size 10 from a population in which the distribution is exponential with mean 1.0. We use (10.5) to form a confidence interval, with approximate confidence level 0.95. For small samples, the true level may be different. Use simulation to find the true level here.

**9.** Suppose we take a random sample of size 10 from a population in which the distribution is exponential with $\lambda = 1$. We use (10.5) to form a confidence interval, with approximate confidence level 0.95. For small samples, the true level may be substantially different. Use simulation to find the true level here. Note: You probably will want to use R's **mean()** function, as well as **sd()** or **var()**. Please note that the latter two use "divide by $n - 1$," requiring an adjustment on your part.

# Part III

# Multivariate Analysis

# Chapter 11

# Multivariate Distributions

Most applications of probability and statistics involve the <u>interaction</u> between variables. For instance, when you buy a book at Amazon.com, the software will likely inform you of other books that people bought in conjunction with the one you selected. Amazon is relying on the fact that sales of certain pairs or groups of books are correlated.

Thus we need the notion of distributions that describe how two or more variables vary together. This chapter develops that notion, **which forms the very core of statistics**, especially in conditional distributions.

## 11.1  Multivariate Distributions: Discrete

Recall that for a single discrete random variable $X$, the distribution of X was defined to be a list of all the values of $X$, together with the probabilities of those values. The same is done for a pair (or more than a pair) of discrete random variables $U$ and $V$.

### 11.1.1  Example: Marbles in a Bag

Suppose we have a bag containing two yellow marbles, three blue ones and four green ones. We choose four marbles from the bag at random, without replacement. Let $Y$ and $B$ denote the number of yellow and blue marbles that we get. Then define the *two-dimensional* pmf of $Y$ and $B$ to be

$$p_{Y,B}(i,j) = P(Y = i \text{ and } B = j) = \frac{\binom{2}{i}\binom{3}{j}\binom{4}{4-i-j}}{\binom{9}{4}} \qquad (11.1)$$

Here is a table displaying all the values of $P(Y = i \text{ and } B = j)$:

| $i \downarrow, j \rightarrow$ | 0 | 1 | 2 | 3 |
|---|---|---|---|---|
| 0 | 0.008 | 0.095 | 0.143 | 0.032 |
| 1 | 0.063 | 0.286 | 0.190 | 0.016 |
| 2 | 0.048 | 0.095 | 0.024 | 0.000 |

So this table is the distribution of the pair $(Y, B)$.

## 11.2  Multivariate Distributions: Continuous

Just as univariate probability density functions are the continuous analog of pmfs, multivariate densities are the continuous analog of joint probability density functions.

### 11.2.1  Motivation and Definition

Extending our previous definition of cdf for a single variable, we define the two-dimensional cdf for a pair of random variables $X$ and $Y$ (discrete or continuous) as

$$F_{X,Y}(u,v) = P(X \le u \text{ and } Y \le v) \qquad (11.2)$$

If $X$ and $Y$ were discrete, we would evaluate that cdf via a double sum of their bivariate pmf. You may have guessed by now that the analog for continuous random variables would be a double integral, and it is. The integrand is the bivariate density:

$$f_{X,Y}(u,v) = \frac{\partial^2}{\partial u\, \partial v} F_{X,Y}(u,v) \qquad (11.3)$$

Densities in higher dimensions are defined similarly.[1]

---

[1] Just as we noted in Section 6.4.2, some random variables are neither discrete nor

As in the univariate case, a bivariate density shows which regions of the X-Y plane occur more frequently, and which occur less frequently.

## 11.2.2   Use of Multivariate Densities in Finding Probabilities and Expected Values

Again by analogy, for any region $A$ in the X-Y plane,

$$P[(X,Y) \in A] = \iint_A f_{X,Y}(u,v) \; du \; dv \qquad (11.4)$$

So, just as probabilities involving a single variable X are found by integrating $f_X$ over the region in question, for probabilities involving X and Y, we take the double integral of $f_{X,Y}$ over that region.

Also, for any function g(X,Y),

$$E[g(X,Y)] = \int_{-\infty}^{\infty} \int_{-\infty}^{\infty} g(u,v) f_{X,Y}(u,v) \; du \; dv \qquad (11.5)$$

where it must be kept in mind that $f_{X,Y}(u,v)$ may be 0 in some regions of the U-V plane. Note that there is no set $A$ here as in (11.4).

Finding marginal densities is also analogous to the discrete case, e.g.,

$$f_X(s) = \int_t f_{X,Y}(s,t) \; dt \qquad (11.6)$$

Other properties and calculations are analogous as well. For instance, the double integral of the density is equal to 1, and so on.

## 11.2.3   Example: Train Rendezvous

Train lines A and B intersect at a certain transfer point, with the schedule stating that trains from both lines will arrive there at 3:00 p.m. However,

---

continuous, there are some pairs of continuous random variables whose cdfs do not have the requisite derivatives. We will not pursue such cases here.

they are often late, by amounts $X$ and $Y$, measured in hours, for the two trains. The bivariate density is

$$f_{X,Y}(s,t) = 2 - s - t, \ 0 < s, t < 1 \tag{11.7}$$

Two friends agree to meet at the transfer point, one taking line A and the other B. Let $W$ denote the time in minutes the person arriving on line B must wait for the friend. Let's find $P(W > 6)$.

First, convert this to a problem involving X and Y, since they are the random variables for which we have a density, and then use (11.4):

$$P(W > 0.1) = P(Y + 0.1 < X) \tag{11.8}$$

$$= \int_{0.1}^{1} \int_{0}^{s-0.1} (2 - s - t) \ dt \ ds \tag{11.9}$$

## 11.3    Measuring Co-variation

### 11.3.1    Covariance

**Definition 22** *The* **covariance** *between random variables* $X$ *and* $Y$ *is defined as*

$$Cov(X, Y) = E[(X - EX)(Y - EY)] \tag{11.10}$$

Suppose that typically when $X$ is larger than its mean, $Y$ is also larger than its mean, and vice versa for below-mean values. Then $(X - EX) \ (Y - EY)$ will usually be positive. In other words, if $X$ and $Y$ are positively correlated (a term we will define formally later but keep intuitive for now), then their covariance is positive. Similarly, if $X$ is often smaller than its mean whenever $Y$ is larger than its mean, the covariance and correlation between them will be negative. All of this is roughly speaking, of course, since it depends on *how much* and *how often* $X$ is larger or smaller than its mean, etc.

There are a number of mailing tubes.

**Linearity in both arguments:**

$$Cov(aX + bY, cU + dV) =$$
$$acCov(X, U) + adCov(X, V) + bcCov(Y, U) + bdCov(Y, V) \tag{11.11}$$

for any constants $a$, $b$, $c$ and $d$.

**Insensitivity to additive constants:**

$$Cov(X, Y + q) = Cov(X, Y) \tag{11.12}$$

for any constant $q$ and so on.

**Covariance of a random variable with itself:**

$$Cov(X, X) = Var(X) \tag{11.13}$$

for any $X$ with finite variance.

**Shortcut calculation of covariance:**

$$Cov(X, Y) = E(XY) - EX \cdot EY \tag{11.14}$$

The proof will help you review some important issues, namely (a) $E(U + V) = EU + EV$, (b) $E(cU) = cEU$ and $Ec = c$ for any constant $c$, and (c) $EX$ and $EY$ are constants in (11.14).

$$
\begin{aligned}
Cov(X, Y) &= E[(X - EX)(Y - EY)] \\
&= E[XY - EX \cdot Y - EY \cdot X + EX \cdot EY] \\
&= E(XY) + E[-EX \cdot Y] + E[-EY \cdot X] + E[EX \cdot EY] \\
&= E(XY) - EX \cdot EY \quad (E[cU] = cEU, \ Ec = c)
\end{aligned}
$$

**Variance of sums:**

$$Var(X + Y) = Var(X) + Var(Y) + 2\,Cov(X, Y) \tag{11.15}$$

This comes from (11.14), the relation $Var(X) = E(X^2) - (EX)^2$ and the corresponding one for Y. Just substitute and do the algebra.

By induction, (11.15) generalizes for more than two variables:

$$Var(W_1 + ... + W_r) = \sum_{i=1}^{r} Var(W_i) + 2 \sum_{1 \leq j < i \leq r} Cov(W_i, W_j) \qquad (11.16)$$

## 11.3.2  Example: The Committee Example Again

Let's find $Var(M)$ in the committee example of Section 4.4.3. In (4.51), we wrote $M$ as a sum of indicator random variables:

$$M = G_1 + G_2 + G_3 + G_4 \qquad (11.17)$$

and found that

$$P(G_i = 1) = \frac{2}{3} \qquad (11.18)$$

for all $i$.

You should review why this value is the same for all $i$, as this reasoning will be used again below. Also review Section 4.4.

From the same reasoning, we know that $(G_i, G_j)$ has the same bivariate distribution for all $i < j$, so the same is true for $Cov(G_i, G_j)$.

Applying (11.16) to (11.17), we have

$$Var(M) = 4\, Var(G_1) + 12\, Cov(G_1.G_2) \qquad (11.19)$$

Finding that first term is easy, from (4.37):

$$Var(G_1) = \frac{2}{3} \cdot \left(1 - \frac{2}{3}\right) = \frac{2}{9} \qquad (11.20)$$

Now, what about $Cov(G_1.G_2)$? Equation (11.14) will be handy here:

$$Cov(G_1.G_2) = E(G_1 G_2) - E(G_1)E(G_2) \qquad (11.21)$$

That first term in (11.21) is

$$
\begin{aligned}
E(G_1 G_2) &= P(G_1 = 1 \text{ and } G_2 = 1) \\
&= P(\text{choose a man on both the first and second pick}) \\
&= \frac{6}{9} \cdot \frac{5}{8} \\
&= \frac{5}{12}
\end{aligned}
$$

That second term in (11.21) is, again from Section 4.4,

$$
\left(\frac{2}{3}\right)^2 = \frac{4}{9} \tag{11.22}
$$

All that's left is to put this together in (11.19), left to the reader.

## 11.4   Correlation

Covariance does measure how much or little $X$ and $Y$ vary together, but it is hard to decide whether a given value of covariance is "large" or not. For instance, if we are measuring lengths in feet and change to inches, then (11.11) shows that the covariance will increase by $12^2 = 144$. Thus it makes sense to scale covariance according to the variables' standard deviations. Accordingly, the *correlation* between two random variables $X$ and $Y$ is defined by

$$
\rho(X,Y) = \frac{Cov(X,Y)}{\sqrt{Var(X)}\sqrt{Var(Y)}} \tag{11.23}
$$

So, correlation is unitless, i.e., does not involve units like feet, pounds, etc.

It can be shown that

- $-1 \leq \rho(X,Y) \leq 1$

- $|\rho(X,Y)| = 1$ if and only if X and Y are exact linear functions of each other, i.e., $Y = cX + d$ for some constants $c$ and $d$.

So the scaling of covariance not only gave a dimensionless (i.e., unitless) quantity, but also one contained within [-1,1]. This helps us recognize what a "large" correlation is, vs. a small one.

### 11.4.1   Sample Estimates

In the statistical context, e.g., Chapter 7, covariance and correlation are population quantities. How can we estimate them using sample values?

As before, we use sample analogs. In the definition, (11.10), think of "E()" as "take the average value in the population." The analog at the sample level would be to take the average value in the sample. Thus define the sample covariance, our sample estimate of $\rho(X,Y)$, as

$$\widehat{Cov}(X,Y) = \frac{1}{n}\sum_{i=1}^{n}(X_i - \overline{X})(Y_i - \overline{Y}) \tag{11.24}$$

for data $(X_1, Y_1), ..., (X_n, Y_n)$.

For correlation, divide by the sample standard deviations:

$$\rho(\widehat{X,Y}) = \frac{\widehat{Cov}(X,Y)}{s_X \, s_Y} \tag{11.25}$$

## 11.5   Sets of Independent Random Variables

Recall from Section 3.3:

**Definition 23** *Random variables X and Y are said to be **independent** if for any sets I and J, the events {X is in I} and {Y is in J} are independent, i.e., P(X is in I and Y is in J) = P(X is in I) P(Y is in J).*

Intuitively, though, it simply means that knowledge of the value of X tells us nothing about the value of Y, and vice versa.

Great mathematical tractability can be achieved by assuming that the $X_i$ in a random vector $X = (X_1, ..., X_k)$ are independent. In many applications, this is a reasonable assumption.

### 11.5.1   Mailing Tubes

In the next few sections, we will look at some commonly-used properties of sets of independent random variables. For simplicity, consider the case $k = 2$, with $X$ and $Y$ being independent (scalar) random variables.

### 11.5.1.1   Expected Values Factor

If $X$ and $Y$ are independent, then

$$E(XY) = E(X)E(Y) \tag{11.26}$$

### 11.5.1.2   Covariance Is 0

If $X$ and $Y$ are independent, we have

$$Cov(X,Y) = 0 \tag{11.27}$$

and thus

$\rho(X,Y) = 0$ as well.

This follows from (11.26) and (11.14).

However, the converse is false. A counterexample is the random pair $(X,Y)$ that is uniformly distributed on the unit disk, $\{(s,t) : s^2 + t^2 \leq 1\}$. Clearly $0 = E(XY) = EX = EY$ due to the symmetry of the distribution about $(0,0)$, so $Cov(X,Y) = 0$ by (11.14).

But $X$ and $Y$ just as clearly are not independent. If for example we know that $X > 0.8$, say, then $Y^2 < 1 - 0.8^2$ and thus $|Y| < 0.6$. If $X$ and $Y$ were independent, knowledge of $X$ should not tell us anything about $Y$, which is not the case here, and thus they are not independent. If we also know that $X$ and $Y$ are bivariate normally distributed (Section 12.1), then zero covariance does imply independence.

### 11.5.1.3   Variances Add

If X and Y are independent, then we have

$$Var(X + Y) = Var(X) + Var(Y). \tag{11.28}$$

This follows from (11.15) and (11.26).

# 11.6   Matrix Formulations

(Note that there is a review of matrix algebra in Appendix B.)

When dealing with multivariate distributions, some very messy equations can be greatly compactified through the use of matrix algebra. We will introduce this here.

Throughout this section, consider a random vector $W = (W_1, ..., W_k)'$ where $'$ denotes matrix transpose, and a vector written horizontally like this without a $'$ means a row vector.

## 11.6.1   Mailing Tubes: Mean Vectors

In statistics, we frequently need to find covariance matrices of linear combinations of random vectors.

**Definition 24** *The expected value of W is defined to be the vector*

$$EW = (EW_1, ..., EW_k)' \tag{11.29}$$

The linearity of the components implies that of the vectors:

For any scalar constants $c$ and $d$, and any random vectors $V$ and $W$, we have

$$E(cV + dW) = cEV + dEW \tag{11.30}$$

where the multiplication and equality is now in the vector sense.

Also, multiplication by a constant matrix factors:

If $A$ is a nonrandom matrix having $k$ columns, then $AW$ is a new random vector, and

$$E(AW) = A\ EW \tag{11.31}$$

## 11.6.2   Covariance Matrices

In moving from random variables, which we dealt with before, to random vectors, we now see that expected value carries over as before. What about variance? The proper extension is the following.

**Definition 25** *The covariance matrix Cov(W) of* $W = (W_1, ..., W_k)'$ *is the* $k \times k$ *matrix whose* $(i, j)^{th}$ *element is* $Cov(W_i, W_j)$.

Note that that and (11.13) imply that the diagonal elements of the matrix are the variances of the $W_i$, and that the matrix is symmetric.

As you can see, in the statistics world, the Cov() notation is "overloaded." If it has two arguments, it is ordinary covariance, between two variables. If it has one argument, it is the covariance matrix, consisting of the covariances of all pairs of components in the argument. When people mean the matrix form, they always say so, i.e., they say "covariance MATRIX" instead of just "covariance."

The covariance matrix is just a way to compactly do operations on ordinary covariances. Here are some important properties:

### 11.6.3   Mailing Tubes: Covariance Matrices

Say $c$ is a constant scalar. Then $cW$ is a $k$-component random vector like $W$, and

$$Cov(cW) = c^2 Cov(W) \tag{11.32}$$

Suppose $V$ and $W$ are independent random vectors, meaning that each component in $V$ is independent of each component of $W$. (But this does NOT mean that the components within $V$ are independent of each other, and similarly for $W$.) Then

$$Cov(V + W) = Cov(V) + Cov(W) \tag{11.33}$$

Of course, this is also true for sums of any (nonrandom) number of independent random vectors.

In analogy with (4.4), for any random vector $Q$,

$$Cov(Q) = E(QQ') - EQ\ (EQ)' \tag{11.34}$$

Suppose $A$ is an $r \times k$ but nonrandom matrix. Then $AW$ is an $r$-component random vector, with its $i^{th}$ element being a linear combination of the elements of $W$. Then one can show that

$$Cov(AW) = A \, Cov(W) \, A' \qquad (11.35)$$

An important special case is that in which $A$ consists of just one row. In this case $AW$ is a vector of length $1$ — a scalar! And its covariance matrix, which is of size $1 \times 1$, is thus simply the variance of that scalar. In other words:

Suppose we have a random vector $U = (U_1, ..., U_k)'$ and are interested in the variance of a linear combination of the elements of U,

$$Y = c_1 U_1 + ... + c_k U_k \qquad (11.36)$$

for a vector of constants $c = (c_1, ..., c_k)'$.

Then

$$Var(Y) = c' Cov(U) c \qquad (11.37)$$

Here are the details: (11.36) is, in matrix terms, $AU$, where $A$ is the one-row matrix consisting of $c'$. Thus (11.35) gives us the right-hand side of (11.36) What about the left-hand side?

In this context, $Y$ is the one-element vector $(Y_1)$. So, its covariance matrix is of size $1 \times 1$, and it sole element is, according to Definition 25, $Cov(Y_1, Y_1)$. But that is $Cov(Y, Y) = Var(Y)$.

## 11.7 Sample Estimate of Covariance Matrix

For a vector-valued random sample $X_1, ..., X_n$,

$$\widehat{Cov}(X) = \sum_{i=1}^{n} X_i X_i' - \overline{X}\,\overline{X}' \qquad (11.38)$$

where

$$\overline{X} = \sum_{i=1}^{n} X_i \qquad (11.39)$$

For instance, say we have data on human height, weight and age. So, $X_1$ is the height, weight and age of the first person in our sample, $X_2$ is the data for the second, and so on.

### 11.7.1 Example: Pima Data

Recall the Pima diabetes data from Section 7.8. For simplicity, let's just look at glucose, blood pressure and insulin. Their population covariance matrix is $3 \times 3$, which we can estimate using R's**scov()** function:

```
> p1 <- pima[,c(2,3,5)]
> cov(p1)
            Gluc         BP       Insul
Gluc  1022.24831   94.43096   1220.9358
BP      94.43096  374.64727    198.3784
Insul 1220.93580  198.37841  13281.1801
```

Or, estimated correlations:

```
> cor(p1)
            Gluc         BP       Insul
Gluc   1.0000000  0.15258959  0.33135711
BP     0.1525896  1.00000000  0.08893338
Insul  0.3313571  0.08893338  1.00000000
```

## 11.8  Mathematical Complements

### 11.8.1  Convolution

**Definition 26** *Suppose g and h are densities of continuous, independent random variables X and Y, respectively. The* **convolution** *of g and h, denoted g * h, is another density, defined to be that of the random variable Z = X + Y. In other words, convolution is a binary operation on the set of all densities.*

258 CHAPTER 11. MULTIVARIATE DISTRIBUTIONS

*If X and Y are nonnegative, then the convolution reduces to*

$$f_Z(t) = \int_0^t g(s)\, h(t-s)\, ds \tag{11.40}$$

You can get intuition on this by considering the discrete case. Say $U$ and $V$ are nonnegative integer-valued random variables, and set $W = U + V$. Let's find $p_W$;

$$
\begin{aligned}
p_W(k) &= P(W = k) \text{ (by definition)} &(11.41)\\
&= P(U + V = k) \text{ (substitution)} &(11.42)\\
&= \sum_{i=0}^{k} P(U = i \text{ and } V = k - i) &(11.43)\\
&= \sum_{i=0}^{k} P(U = i)\, P(V = k - i) &(11.44)\\
&= \sum_{i=0}^{k} p_U(i) p_V(k - i) &(11.45)
\end{aligned}
$$

Review the analogy between densities and pmfs in our unit on continuous random variables, Section 6.5, and then see how (11.40) is analogous to (11.41) through (11.45):

- $k$ in (11.41) is analogous to $t$ in (11.40)

- the limits 0 to $k$ in (11.45) are analogous to the limits 0 to $t$ in (11.40)

- the expression $k - i$ in (11.45) is analogous to $t - s$ in (11.40)

- and so on

### 11.8.1.1 Example: Backup Battery

Suppose we have a portable machine that has compartments for two batteries. The main battery has lifetime $X$ with mean 2.0 hours, and the backup's lifetime $Y$ has mean life 1 hour. One replaces the first by the second as soon as the first fails. The lifetimes of the batteries are exponentially distributed and independent. Let's find the density of $W$, the time that the system is operational (i.e., the sum of the lifetimes of the two batteries).

Recall that if the two batteries had the same mean lifetimes, $W$ would have a gamma distribution. That's not the case here, but we notice that the distribution of $W$ is a convolution of two exponential densities, as it is the sum of two nonnegative independent random variables. Using (11.40), we have

$$f_W(t) = \int_0^t f_X(s)f_Y(t-s)\, ds = \int_0^t 0.5e^{-0.5s}e^{-(t-s)}\, ds = e^{-0.5t} - e^{-t},$$
$$0 < t < \infty$$

$$(11.46)$$

## 11.8.2 Transform Methods

We often use the idea of *transform* functions. For example, you may have seen *Laplace transforms* in a math or engineering course. The functions we will see here differ from this by just a change of variable.

This technique will be used here to show that if $X$ and $Y$ are independent, Poisson-distributed random variables, their sum again has a Poisson distribution.

### 11.8.2.1 Generating Functions

Here we will discuss one of the transforms, the *generating function*. For any nonnegative-integer valued random variable $V$, its generating function is defined by

$$g_V(s) = E(s^V) = \sum_{i=0}^{\infty} s^i p_V(i),\ 0 \le s \le 1 \qquad (11.47)$$

For instance, suppose N has a geometric distribution with parameter $p$, so that $p_N(i) = (1-p)^{i-1}\, p,\ i = 1, 2, \ldots$ Then

$$g_N(s) = \sum_{i=1}^{\infty} s^i \cdot (1-p)^{i-1} p \qquad (11.48)$$

$$= \frac{p}{1-p} \sum_{i=1}^{\infty} s^i \cdot (1-p)^i \qquad (11.49)$$

$$= \frac{p}{1-p} \frac{(1-p)s}{1-(1-p)s} \qquad (11.50)$$

$$= \frac{ps}{1-(1-p)s} \qquad (11.51)$$

Why restrict s to the interval [0,1]? The answer is that for $s > 1$ the series in (11.47) may not converge. for $0 \le s \le 1$, the series does converge. To see this, note that if s = 1, we just get the sum of all probabilities, which is 1.0. If a nonnegative s is less than 1, then $s^i$ will also be less than 1, so we still have convergence.

One use of the generating function is, as its name implies, to generate the probabilities of values for the random variable in question. In other words, if you have the generating function but not the probabilities, you can obtain the probabilities from the function. Here's why: For clarify, write (11.47) as

$$g_V(s) = P(V=0) + sP(V=1) + s^2 P(V=2) + ... \qquad (11.52)$$

Plugging $s = 0$ into this equation, we see that

$$g_V(0) = P(V=0) \qquad (11.53)$$

So, we can obtain $P(V=0)$ from the generating function. Now differentiating (11.47) with respect to s,[2] we have

$$g_V'(s) = \frac{d}{ds} \left[ P(V=0) + sP(V=1) + s^2 P(V=2) + ... \right]$$
$$= P(V=1) + 2sP(V=2) + ... \qquad (11.54)$$

So, we can obtain $P(V = 1)$ from $g_V'(0)$, and in a similar manner can calculate the other probabilities from the higher derivatives.

---

[2]Here and below, to be mathmmatically rigorous, we would need to justify interchanging the order of summation and differentiatiation.

Note too:

$$g'_V(s) = \frac{d}{ds} E(s^V) = E(Vs^{V-1}) \qquad (11.55)$$

So,

$$g'_V(1) = EV \qquad (11.56)$$

In other words, we can use the generating function to find the mean as well.

Also, if $X$ and $Y$ are independent, then $g_{X+Y} = g_X g_Y$. (Exercise 11.)

### 11.8.2.2 Sums of Independent Poisson Random Variables Are Poisson Distributed

Suppose packets come in to a network node from two independent links, with counts $N_1$ and $N_2$, Poisson distributed with means $\mu_1$ and $\mu_2$. Let's find the distribution of $N = N_1 + N_2$, using a transform approach.

We first need to find the Poisson generating function, say for a Poisson random variable $M$ with mean $\lambda$:

$$g_M(s) = \sum_{i=0}^{\infty} s^i \frac{e^{-\lambda} \lambda^i}{i!} = e^{-\lambda + \lambda s} \sum_{i=0}^{\infty} \frac{e^{-\lambda s} (\lambda s)^i}{i!} \qquad (11.57)$$

But the summand is the pmf for a Poisson distribution with mean $\lambda s$, and thus the sum is 1.0. In other words

$$g_M(s) = e^{-\lambda + \lambda s} \qquad (11.58)$$

So we have

$$g_N(s) = g_{N_1}(s) \, g_{N_2}(s) = e^{-\nu + \nu s} \qquad (11.59)$$

where $\nu = \mu_1 + \mu_2$.

But the last expression in (11.59) is the generating function for a Poisson distribution too! And since there is a one-to-one correspondence between distributions and transforms, we can conclude that $N$ has a Poisson distribution with parameter $\nu$. We of course knew that $N$ would have mean $\nu$ but did not know that $N$ would have a Poisson distribution.

So: A sum of two independent Poisson variables itself has a Poisson distribution. By induction, this is also true for sums of $k$ independent Poisson variables.

## 11.9   Exercises

**Mathematical problems:**

**1.** Let $X$ and $Y$ denote the number of dots we get when we roll a pair of dice. Find $\rho(X, S)$, where $S = X + Y$.

**2.** In the marbles example, Section 11.1.1, find $Cov(Y, B)$.

**3.** Consider the toy population example, Section 7.3.1. Suppose we take a sample of size 2 *without* replacement. Find $Cov(X_1, X_2)$.

**4.** Suppose $(X, Y)$ has the density

$$f_{X,Y}(s,t) = 8st, \ 0 < t < s < 1 \tag{11.60}$$

Find $P(X + Y > 1)$, $f_Y(t)$ and $\rho(X, Y)$.

**5.** Say $X$ and $Y$ are independent with density $2t$ on $(0, 1)$, 0 elsewhere. Find $f_{X+Y}(t)$.

**6.** Using (11.58), verify that the values of $P(M = 0)$, $P(M = 1)$ and $EM$ calculated from the generating function are indeed correct.

**7.** In the parking place example, Section 5.4.1.2, find $Cov(D, N)$. Hints: You'll need to evaluate expressions like (5.14), and do some algebra. Be presistent!

**8.** Consider a random variable $X$ having negative binomial distribution with $r = 3$ and $p = 0.4$. Find the *skewness* of this distribution, $E[((X - \mu)/\sigma)^3$, where $\mu$ and $\sigma$ are the mean and standard deviation, respectively. Hint: Recall that a negative binomial random variable can be written as a sum of independent geometric random variables. You'll need to review Section 5.4.1, and have some persistence!

**9.** Say the random variable $X$ is *categorical*, taking on the values $1, 2, ..., c$, with $i$ indicating category $i$. For instance, $X$ might be a telephone country code for a company chosen at random from the world, such as 1 for the US and 852 for Hong Kong. Let $p_i = P(X = i)$, $i = 1, ..., c$. Say we have a random sample $X_1, ..., X_n$ from this distribution, with $N_j$ denoting the number of $X_i$ that are equal to $j$. Our estimates of the $p_j$ are $\widehat{p}_j = \frac{N_j}{n}$.

Expressing in terms of the $p_j$ and $n$, show that the covariance matrix of $\widehat{p} = (\widehat{p}_1, ..., \widehat{p}_c)'$ has its $(i, j)$ element equal to

$$
\Sigma_{ij} = \begin{cases} -np_i p_j, & i \neq j \\ np_i(1 - p_i), & i = j \end{cases} \tag{11.61}
$$

Hint: Define analogs $W_1, ..., W_n$ of the Bernoulli variables in Section 5.4.2, each a vector of length $c$, as follows. Say $X_i = j$; then define $W_i$ to be equal to 1 in component $j$ and 0 elsewhere. In other words, for each fixed $j$, the scalars $W_{1j}, ..., W_{nj}$ are indicator variables for category $j$. Now, apply (11.30) and (11.33) to

$$
T = \sum_{i=1}^{n} W_i \tag{11.62}
$$

and use the fact that the $W_{ij}$ are indicators.

**10**. Use the results of Problem 9 to find

$$
Var(\widehat{F}_X(t_1) - \widehat{F}_X(t_2)) \tag{11.63}
$$

in (8.24). Your expression should involve $F_X(t_i)$, $i = 1, 2$.

**11**. Suppose $X$ and $Y$ are nonnegative integer-valued random variables. Show that $g_{X+Y} = g_X g_Y$.

**12**. Give an alternate derivation of the result of Section 11.8.2.2, using the convolution computations in (11.41).

**Computational and data problems:**

**13**. In the BMI data (Section 7.8), find the sample estimate of the population correlation between BMI and blood pressure.

**14**. Consider the Pima data, Section 7.8. It was found that the dataset contains a number of outliers/errors (Section 7.12.2). There we looked at the data columns individually, but there are ways to do this jointly, such as *Mahalanobis distance*, defined as follows.

For a random vector $W$ with mean $\mu$ and covariance matrix $\Sigma$. the distance is

$$d = (W - \mu)'\Sigma^{-1}(W - \mu) \qquad (11.64)$$

(Technically we should use $\sqrt{d}$, but we generally skip this.) Note that this is a random quantity, since $W$ is random.

Now suppose we have a random sample $W_1, ..., W_n$ from this distribution. For each data point, we can find

$$d_i = (W_i - \widehat{\mu})'\widehat{\Sigma}^{-1}(W_i - \widehat{\mu}) \qquad (11.65)$$

Any data points with suspiciously high values of this distance are then considered for possible errors and so on.

Apply this idea to the Pima data, and see whether most of the points with large distance have invalid 0s.

**15**. Obtain the dataset **prgeng** from my **freqparcoord** package on CRAN. Find the sample correlation between age and income for men, then do the same for women. And comment: Do you think the difference is reflected in the population? If so, why? [Unfortunately, there are no handy formulas for standard errors of $\widehat{\rho}$.]

# Chapter 12

# The Multivariate Normal Family of Distributions

Intuitively, this family has densities which are shaped like multidimensional bells, just like the univariate normal has the famous one-dimensional bell shape.

## 12.1 Densities

Let's look at the bivariate case first. The joint distribution of $X$ and $Y$ is said to be *bivariate normal* if their density is

$$
f_{X,Y}(s,t) = \frac{1}{2\pi\sigma_1\sigma_2\sqrt{1-\rho^2}} e^{-\frac{1}{2(1-\rho^2)}\left[\frac{(s-\mu_1)^2}{\sigma_1^2} + \frac{(t-\mu_2)^2}{\sigma_2^2} - \frac{2\rho(s-\mu_1)(t-\mu_2)}{\sigma_1\sigma_2}\right]},
$$

$$
-\infty < s, t < \infty
$$

(12.1)

This is pictured on the cover of this book, and the two-dimensional bell shape has a pleasing appearance. But the expression for the density above looks horrible, and it is. Don't worry, though, as we won't work with this directly. It's important for conceptual reasons, as follows.

First, note the parameters here: $\mu_1$, $\mu_2$, $\sigma_1$ and $\sigma_2$ are the means and standard deviations of $X$ and $Y$, while $\rho$ is the correlation between $X$ and

$Y$. So, we have a five-parameter family of distributions.

More generally, the multivariate normal family of distributions is parameterized by one vector-valued quantity, the mean $\mu$, and one matrix-valued quantity, the covariance matrix $\Sigma$. Specifically, suppose the random vector $X = (X_1, ..., X_k)'$ has a $k$-variate normal distribution. Then the density has this form:

$$f_X(t) = ce^{-0.5(t-\mu)'\Sigma^{-1}(t-\mu)} \tag{12.2}$$

Here $c$ is a constant, needed to make the density integrate to 1.0. It turns out that

$$c = \frac{1}{(2\pi)^{k/2}\sqrt{det(\Sigma)}} \tag{12.3}$$

but we'll never use this fact.

Here again $'$ denotes matrix transpose, -1 denotes matrix inversion and $det()$ means determinant. Again, note that $t$ is a $k \times 1$ vector.

Since the matrix is symmetric, there are $k(k + 1)/2$ distinct parameters there, and $k$ parameters in the mean vector, for a total of $k(k + 3)/2$ parameters for this family of distributions.

## 12.2   Geometric Interpretation

Now, let's look at some pictures, generated by R code which I've adapted from one of the entries in the old R Graph Gallery, now sadly defunct. Both are graphs of bivariate normal densities, with $EX_1 = EX_2 = 0$, $Var(X_1) = 10$, $Var(X_2) = 15$ and a varying value of the correlation $\rho$ between $X_1$ and $X_2$. Figure 12.1 is for the case $\rho = 0.2$.

The surface is bell-shaped, though now in two dimensions instead of one. Again, the height of the surface at any $(s, t)$ point the relative likelihood of $X_1$ being near $s$ and $X_2$ being near $t$. Say for instance that $X_1$ is height and $X_2$ is weight. If the surface is high near, say, (70,150) (for height of 70 inches and weight of 150 pounds), it means that there are a lot of people whose height and weight are near those values. If the surface is rather low there, then there are rather few people whose height and weight are near those values.

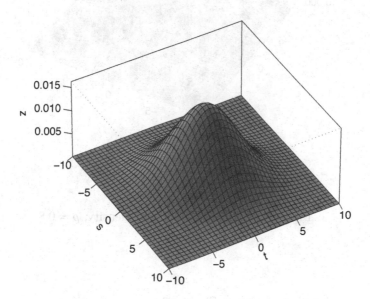

Figure 12.1: Bivariate Normal Density, $\rho = 0.2$

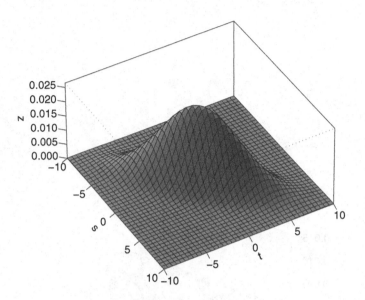

Figure 12.2: Bivariate Normal Density, $\rho = 0.8$

Now compare that picture to Figure 12.2, with $\rho = 0.8$.

Again we see a bell shape, but in this case "narrower." In fact, you can see that when $X_1$ (i.e., $s$) is large, $X_2$ ($t$) tends to be large too, and the same for "large" replaced by small. By contrast, the surface near (5,5) is much higher than near (5,-5), showing that the random vector $(X_1, X_2)$ is near (5,5) much more often than (5,-5).

All of this reflects the high correlation (0.8) between the two variables. If we were to continue to increase $\rho$ toward 1.0, we would see the bell become narrower and narrower, with $X_1$ and $X_2$ coming closer and closer to a linear relationship, one which can be shown to be

$$X_1 - \mu_1 = \frac{\sigma_1}{\sigma_2}(X_2 - \mu_2) \qquad (12.4)$$

In this case, that would be

$$X_1 = \sqrt{\frac{10}{15}} X_2 = 0.82 X_2 \qquad (12.5)$$

## 12.3 R Functions

R provides functions that compute probabilities involving this family of distributions, in the library **mvtnorm**. Of particular interest is the R function **pmvnorm()**, which computes probabilities of "rectangular" regions for multivariate normally distributed random vectors. The arguments we'll use for this function here are:

- **mean**: the mean vector

- **sigma**: the covariance matrix

- **lower, upper**: bounds for a multidimensional "rectangular" region of interest

Since a multivariate normal distribution is characterized by its mean vector and covariance matrix, the first two arguments above shouldn't suprise you. But what about the other two?

The function finds the probability of our random vector falling into a multidimensional rectangular region that we specify, through the arguments are **lower** and **upper**. For example, suppose we have a trivariate normally distributed random vector $(U, V, W)'$, and we want to find

$$P(1.2 < U < 5 \text{ and } -2.2 < V < 3 \text{ and } 1 < W < 10) \qquad (12.6)$$

Then **lower** would be (1.2,-2.2,1) and **upper** would be (5,3,10).

Note that these will typically be specified via R's **c()** function, but default values are recycled versions of **-Inf** and **Inf**, built-in R constants for $-\infty$ and $\infty$.

An important special case is that in which we specify **upper** but allow **lower** to be the default values, thus computing a probability of the form

$$P(W_1 \le c_1, ..., W_r \le c_r) \qquad (12.7)$$

The same library contains **rmvnorm()**, to generate multivariate normally distributed random numbers. The call

```
rmvnorm(n,mean,sigma)
```

generates **n** random vectors from the multivariate normal distribution specified by **mean** and **sigma**.

## 12.4    Special Case: New Variable Is a Single Linear Combination of a Random Vector

Suppose the vector $U = (U_1, ..., U_k)'$ has a $k$-variate normal distribution, and we form the scalar

$$Y = c_1U_1 + ... + c_kU_k \tag{12.8}$$

Then Y is univariate normal, and its (exact) variance is given by (11.37). Its mean is obtained via (11.31).

We can then use the R functions for the univariate normal distribution, e.g., **pnorm()**.

## 12.5    Properties of Multivariate Normal Distributions

**Theorem 27** *Suppose $X = (X_1, ..., X_k)'$ has a multivariate normal distribution with mean vector $\mu$ and covariance matrix $\Sigma$. Then:*

   (a) *The contours of $f_X$ are k-dimensional ellipsoids.  In the case $k = 2$ for instance, where we can visualize the density of $X$ as a three-dimensional surface, the contours, or level sets, for points at which the bell has the same height (think of a topographical map) are elliptical in shape. The larger the correlation (in absolute value) between $X_1$ and $X_2$, the more elongated the ellipse. When the absolute correlation reaches 1, the ellipse degenerates into a straight line.*

(b) Let $A$ be a constant (i.e., nonrandom) matrix with $k$ columns. Then the random vector $Y = AX$ also has a multivariate normal distribution.[1]

The parameters of this new normal distribution must be $EY = A\mu$ and $Cov(Y) = A\Sigma A'$, by (11.31) and (11.35).

(c) If $U_1, ..., U_m$ are each univariate normal and they are independent, then they jointly have a multivariate normal distribution. (In general, though, having a normal distribution for each $U_i$ does not imply that they are jointly multivariate normal.)

(d) Suppose $W$ has a multivariate normal distribution. The conditional distribution of some components of $W$, given other components, is again multivariate normal.

Part [(b)] has some important implications:

(i) The lower-dimensional marginal distributions are also multivariate normal. For example, if $k = 3$, the pair $(X_1, X_3)'$ has a bivariate normal distribution, as can be seen by setting

$$A = \begin{pmatrix} 1 & 0 & 0 \\ 0 & 0 & 1 \end{pmatrix} \qquad (12.9)$$

in (b) above.

(ii) Scalar linear combinations of X are normal. In other words, for constant scalars $a_1, ..., a_k$, set $a = (a_1, ..., a_k)'$. Then the quantity $Y = a_1 X_1 + ... + a_k X_k$ has a univariate normal distribution with mean $a'\mu$ and variance $a'\Sigma a$.

(iii) Vector linear combinations are multivariate normal. Again using the case k = 3 as our example, consider $(U, V)' = (X_1 - X_3, X_2 - X_3)$. Then set

$$A = \begin{pmatrix} 1 & 0 & -1 \\ 0 & 1 & -1 \end{pmatrix} \qquad (12.10)$$

(iv) The r-component random vector X has a multivariate normal distribution if and only if $c'X$ has a univariate normal distribution for all constant r-component vectors c.

---

[1] Note that this is a generalization of the material on affine transformations on page 198.

# 12.6    The Multivariate Central Limit Theorem

The multidimensional version of the Central Limit Theorem holds. A sum of independent identically distributed (*iid*) random vectors has an approximate multivariate normal distribution. Here is the theorem:

**Theorem 28** *Suppose $X_1, X_2, \ldots$ are independent random vectors, all having the same distribution which has mean vector $\mu$ and covariance matrix $\Sigma$. Form the new random vector $T = X_1 + \ldots + X_n$. Then for large $n$, the distribution of $T$ is approximately multivariate normal with mean $n\mu$ and covariance matrix $n\Sigma$.*

For example, since a person's body consists of many different components, the CLT (a non-independent, non-identically version of it) explains intuitively why heights and weights are approximately bivariate normal. Histograms of heights will look approximately bell-shaped, and the same is true for weights. The multivariate CLT says that three-dimensional histograms—plotting frequency along the "Z" axis against height and weight along the "X" and "Y" axes—will be approximately three-dimensional bell-shaped.

The proof of the multivariate CLT is easy, from Property (iv) above. Say we have a sum of iid random vectors:

$$S = X_1 + \ldots + X_n \tag{12.11}$$

Then

$$c'S = c'X_1 + \ldots + c'X_n \tag{12.12}$$

Now on the right side we have a sum of iid *scalars*, not vectors, so the univariate CLT applies! We thus know the right-hand side is a approximately normal for all c, which means $c'S$ is also approximately normal for all c, which then by (iv) above means that S itself is approximately multivariate normal.

## 12.7 Exercises

**Mathematical problems:**

**1.** In Section 11.5.1.2, it was shown that 0 correlation does not imply independence. It was mentioned, though, that this implication does hold if the two random variables involved have a bivariate nromal distribution. Show that this is true, using the fact that $X$ and $Y$ are independent if and only if $f_{X,Y} = f_X \cdot f_Y$.

**2.** Consider the context of Problem 9, Chapter 11. Say we toss a die 600 times. Let $N_j$ denote the number of times the die shows $j$ dots. Citing the Multivariate Central Limit Theorem, find the approximate value of $P(N_1 > 100, N_2 < 100)$. Hint: Use the "dummy vector" approach of Problem 9, Chapter 11.

**3.** Suppose a length-$p$ random vector $W$ having a multivariate normal distribution with mean $\mu$ and covariance matrix $\Sigma$. As long as this distribution is not degenerate, i.e., one component of $W$ is not an exact linear combination of the others, then $\Sigma^{-1}$ will exist. Consider the random variable

$$Z = (W - \mu)'\Sigma^{-1}(W - \mu) \qquad (12.13)$$

Show that $Z$ has a chi-square distribution with $p$ degrees of freedom. Hint: Use the PCA decomposition of $\Sigma$.

**Computational and data problems:**

**4.** Consider the Pima data, Section 7.8. Explore how well a bivariate normal distribution fits this pair of random variables. Estimate $\mu$ and $\Sigma$ from the data, then use **pmvnorm()** (Section 12.3) to estimate $P(\text{BMI} < c, \text{b.p.} < d)$ for various values of $c$ and $d$, assuming normality. Compare these estimated probabilities to the corresponding sample proportions. ·

**5.** To further illustrate the effect of increasing correlation, generate $(X_1, X_2)'$ pairs from a bivariate normal with mean $(0,0)'$ and covariance matrix

$$\begin{pmatrix} 1 & \rho \\ \rho & 1 \end{pmatrix} \qquad (12.14)$$

Through this simulation, estimate $P(|X_1 - X_2| < 0.1)$, for various values of $\rho$.

# Chapter 13

# Mixture Distributions

One of R's built-in datasets is **Faithful**, data on the Old Faithful geyser in America's Yellowstone National Park. Let's plot the waiting times between eruptions:

```
> plot(density(faithful$waiting))
```

The result is shown in Figure 13.1. The density appears to be *bimodal*, i.e., to have two peaks. In fact, it looks like the density somehow "combines" two normal densities. In this chapter, we'll explore this idea, termed *mixture distributions*.

Mixture distributions are everywhere, as there are almost always major subpopulations one might account for. If $H$ is human height and $G$ is gender, we may again find the distribution of $H$ is bimodal; the individual conditional densities of $H \mid G =$ male and $H \mid G =$ female may each be bell-shaped but the unconditional density of $H$ might be bimodal.[1]

A lot of the usefulness of the mixture distributions notion stems from its use in discovering interesting aspects of our data. In the Old Faithful data, for instance, the bimodal shape is intriguing. Are there two physical processes at work deep underground?

As always, much of our analysis is based on expected value and variance. But this can be subtle in the case of mixture distributions. For instance, in the height example above, suppose the population mean height of men is 70.2 inches, with variance 5.8, with the values for women being 68.8 and

---

[1] Note of course that we could define other subpopulations besides gender, or define them in combination.

**density.default(x = faithful$waiting)**

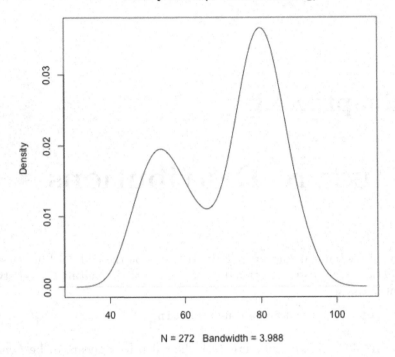

Figure 13.1: Old Faithful Waiting times

5.4. What are the unconditional mean and variance of height?

You might guess that the mean is (70.2 + 68.8) / 2 = 69.5 and the variance is (5.8 + 5.4) / 2 = 5.6. That first guess would be correct, but the second is wrong. To see this, we will first need to develop some probability infrastructure, called *iterated expectations*. We'll then apply it to mixture distributions (though it also has many other applications).

# 13.1   Iterated Expectations

This section has an abstract title, but the contents are quite useful.

## 13.1.1    Conditional Distributions

The very core of predictive methods — including the much vaunted *machine learning* methods in the press so much these days — is something quite basic: conditional probability. For instance, in a medical context, we wish to know the probability that a patient has s certain disease, *given* her test results. Similar notions come up in marketing (Will this Web user click on this icon, given his demographics and past click history?), finance (How much will bond prices rise, given recent prices?) and so on.

Generalizing, the next step after conditional probability is that of a conditional distribution. Just as we can define bivariate pmfs, we can also speak of conditional pmfs. Suppose we have random variables $U$ and $V$. Then for example the quantities $P(U = i \mid V = 5)$, as we vary $i$, form the conditional pmf of $U$, given $V = 5$.

We can then talk about expected values within such pmfs. In our bus ridership example (Section 1.1), for instance, we can talk about

$$E(L_2 \mid B_1 = 0) \qquad (13.1)$$

In notebook terms, think of many replications of watching the bus during times 1 and 2. Then (13.1) is defined to be the long-run average of values in the $L_2$ column, **among those rows** in which the $B_1$ column is 0. (And by the way, make sure you understand why (13.1) works out to be 0.6.)

## 13.1.2    The Theorem

The key relation says, in essence,

> The overall mean of $V$ is a weighted average of the conditional means of $V$ given $U$. The weights are the pmf of $U$.

Note again that $E(V \mid U = c)$ is defined in "notebook" terms as the long-run average of $V$, *among those lines in which $U = c$.*

Here is the formal version:

Suppose we have random variables $U$ and $V$, with $U$ discrete and with $V$

having an expected value. Then

$$E(V) = \sum_c P(U = c)\, E(V \mid U = c) \qquad (13.2)$$

where $c$ ranges through the support of $U$.

So, just as $EX$ was a weighted average in (3.19), with weights being probabilities, we see here that the unconditional mean is a weighted average of the conditional mean.

In spite of its intimidating form, (13.2) makes good intuitive sense, as follows: Suppose we want to find the average height of all students at a university. Each department measures the heights of its majors, then reports the mean height among them. Then (13.2) says that to get the overall mean in the entire school, we should take a *weighted* average of all the within-department means, with the weights being the proportions of each department's student numbers among the entire school. Clearly, we would not want to take an unweighted average, as that would count tiny departments just as much as large majors.

Here is the derivation (reader: supply the reasons!).

$$
\begin{aligned}
EV \;&=\; \sum_d d\, P(V = d) & (13.3) \\[4pt]
&=\; \sum_d d \sum_c P(U = c \text{ and } V = d) & (13.4) \\[4pt]
&=\; \sum_d d \sum_c P(U = c)\, P(V = d \mid U = c) & (13.5) \\[4pt]
&=\; \sum_d \sum_c d\, P(U = c)\, P(V = d \mid U = c) & (13.6) \\[4pt]
&=\; \sum_c \sum_d d\, P(U = c)\, P(V = d \mid U = c) & (13.7) \\[4pt]
&=\; \sum_c P(U = c) \sum_d d\, P(V = d \mid U = c) & (13.8) \\[4pt]
&=\; \sum_c P(U = c)\, E(V \mid U = c) & (13.9)
\end{aligned}
$$

There is also a continuous version:

$$E(W) = \int_{-\infty}^{\infty} f_V(t) \, E(W \mid V = t) \, dt \qquad (13.10)$$

## 13.1.3 Example: Flipping Coins with Bonuses

A game involves flipping a coin $k$ times. Each time you get a head, you get a bonus flip, not counted among the $k$. (But if you get a head from a bonus flip, that does not give you its own bonus flip.) Let $X$ denote the number of heads you get among all flips, bonus or not. Let's find the expected value of $X$.

We should be careful not to come to hasty conclusions. The situation here "sounds" binomial, but $X$, based on a variable number of trials, doesn't fit the definition of binomial. But let $Y$ denote the number of heads you obtain through nonbonus flips. $Y$ then has a binomial distribution with parameters $k$ and 0.5. To find the expected value of $X$, we'll condition on $Y$.

The principle of iterated expectation can easily get us $EX$:

$$
\begin{aligned}
EX &= \sum_{i=1}^{k} P(Y = i) \, E(X|Y = i) & (13.11) \\
&= \sum_{i=1}^{k} P(Y = i) \, (i + 0.5i) & (13.12) \\
&= 1.5 \sum_{i=1}^{k} P(Y = i) \, i & (13.13) \\
&= 1.5 \, EY & (13.14) \\
&= 1.5 \cdot k/2 & (13.15) \\
&= 0.75k & (13.16)
\end{aligned}
$$

To understand that second equality, note that if $Y = i$, $X$ will already include those $i$ heads, the nonbonus heads, and since there will be $i$ bonus flips, the expected value of the number of bonus heads will be $0.5i$.

The reader should ponder how one might solve this problem without iterated expectation. It would get quite complex.

### 13.1.4    Conditional Expectation as a Random Variable

Here we derive a more famous version of (13.2). It will seem more abstract, but it is quite useful, e.g., in our discussion of mixture distributions below.

Consider the context of (13.2). Define a new random variable as follows. First, define $g(c) = E(V|U = c)$. Then $g()$ is an ordinary "algebra/calculus-style" function, so we can apply (3.34). In those terms, (13.2) says

$$EV = E[g(U)] \tag{13.17}$$

Finally, define a new random variable, denoted $E(V \mid U)$, as

$$E(V \mid U) = g(U) \tag{13.18}$$

Let's make this concrete. Say we flip a coin, defining $M$ to be 1 for heads, 2 for tails. Then we roll $M$ dice, yielding a total of $X$ dots. Then

$$g(1) = 3.5, \ g(2) = 7 \tag{13.19}$$

The random variable $E(X \mid M)$ then takes on the values 3.5 and 7, with probabilities 0.5 and 0.5.

So, we have that

$$EV = E[E(V \mid U)] \tag{13.20}$$

The difference between this version and the earlier one is more one of notation than substance, but it makes things much easier, as we will see below.

### 13.1.5    What about Variance?

Equation (13.20) can be used to show a variance version of (13.2):

$$Var(V) = E[Var(V \mid U)] + Var[E(V \mid U)] \tag{13.21}$$

This may seem counterintuitive. That first term seems to be a plausible analog of (13.20), but why the second? Think back to the discussion following (13.2). The overall variance in heights at that university should involve the average of within-department variation, yes, but it should also take into account variation from one department to another, hence the second term.

# 13.2  A Closer Look at Mixture Distributions

We have a random variable $X$ of interest whose distribution depends on which subpopulation we are in. Let $M$ denote the ID number of the subpopulation. In the cases considered here, $M$ is discrete, i.e., the number of subpopulations is finite.

Consider a study of adults, with $H$ and $G$ denoting height and gender, respectively. Pay close attention to whether we are looking at an unconditional or conditional distribution! For instance, $P(H > 73)$ is the overall popular proportion of people taller than 73 inches, while $P(H > 73 \mid G = male)$ is the proportion for men.

We say that the distribution of $H$ is a *mixture* of two distributions, in this case the two within-gender distributions, $f_{H|G=male}$ and $f_{H|G=female}$. Then, assuming the two subpopulations are of the same size, i.e., $P(male) = 0.5$,

$$f_H(t) = 0.5 \, f_{H|G=male}(t) + 0.5 \, f_{H|G=female}(t) \tag{13.22}$$

We say the mixture here has two *components*, due to there being two subpopulations. In general, write the proportions of the $r$ subpopulations (0.5 and 0.5 above) as $p_1, ..., p_r$, with corresponding conditional densities $f_1, ..., f_r$.

Going back to our general notation, let $X$ be a random variable having a mixture distribution, with subpopulation ID denoted by $M$. Note that $X$ could be vector-valued, e.g., the pair (height, weight).

## 13.2.1  Derivation of Mean and Variance

Denote the means and variances of $X$ in subpopulation $i$ as

$$\mu_i = E(X \mid M = i) \tag{13.23}$$

and

$$\sigma_i^2 = Var(X \mid M = i) \tag{13.24}$$

It is often necessary to know the (unconditional) mean and variance of $X$, in terms of the $\mu_i$ and $\sigma_i^2$. Here is a good chance to apply our formulas above for iterated means and variances. First, the mean, using (13.20):

$$EX = E[E(X \mid M)] \tag{13.25}$$

How do we evaluate the right-hand side? The key point is that $E(X \mid M)$ is a discrete random variable that takes on the values $\mu_1, ..., \mu_r$, with probabilities $p_1, ..., p_r$. So, the expected value of this random variable is an easy application of (3.19):

$$E[\, E(X \mid M)\, ] = \sum_{i=1}^{r} p_i \mu_i \tag{13.26}$$

Thus we now have the (unconditional) mean of $X$

$$EX = \sum_{i=1}^{r} p_i \mu_i = \mu \tag{13.27}$$

Finding the variance of $X$ is almost as easy, just a little more care needed. From (13.21),

$$Var(X) = E[Var(X \mid M)] + Var[E(X \mid M)] \tag{13.28}$$

Let's first look at that second term on the right. Again, keep in mind that $E(X \mid M)$ is a discrete random variable, taking on the values $\mu_1, ..., \mu_r$, with probabilities $p_1, ..., p_r$. That second term asks for the variance of that random variable, which from (4.1) is

$$\sum_{i=1}^{r} p_i \, (\mu_i - \mu)^2 \tag{13.29}$$

Now consider the first term on the right. We have

$$Var(X \mid M) = \sigma_M^2 \tag{13.30}$$

so we are taking the expected value (outer operation in that first term) of a random variable which takes on the values $\sigma_1^2,..., \sigma_r^2$, with probabilities $p_1,...,p_r$. Thus

$$E[Var(X \mid M)] = \sum_{i=1}^{r} p_i \, \sigma_i^2 \tag{13.31}$$

So, we're done!

$$Var(X) = \sum_{i=1}^{r} p_i \, \sigma_i^2 + \sum_{i=1}^{r} p_i \, (\mu_i - \mu)^2 \tag{13.32}$$

Again, the interpretation is that the overall population variance is the weighted average of the subpopulation variances, plus a term accounting for the variation among subpopulations of the mean.

## 13.2.2   Estimation of Parameters

How do we use our sample data to estimate the parameters of a mixture distribution? Say for instance $r = 3$ and $p_{X|m=i}$ is Poisson with mean $q_i$. That's a total of six parameters; how do we estimate them from our data $X_1,...,X_n$?

There are the usual methods, of course, MLE and MM. We might try this, but due to convergence worries, a more sophisticated method, the *Expectation-Maximization (EM) algorithm*, is more commonly used. It is not guaranteed to converge either, but it generally is successful.

Unfortunately, the theory behind EM is too complex for convenient exposition in this book. But we'll use it:

### 13.2.2.1   Example: Old Faithful Estimation

One of the R packages for EM is **mixtools**. Let's see how to fit a model of mixed normals:

```
> library(mixtools)
> mixout <- normalmixEM(faithful$waiting,lambda=0.5,
    mu=c(55,80),sigma=10,k=2)
> str(mixout)
```

```
List of 9
 $ x          : num  [1:272] 79 54 74 62 85 55 88
     85 51 85 ...
 $ lambda     : num  [1:2] 0.361 0.639
 $ mu         : num  [1:2] 54.6 80.1
 $ sigma      : num  [1:2] 5.87 5.87
 $ loglik     : num  -1034
 ...
```

Here $\lambda$ is the vector of the $p_i$. In the call, we set the initial guess to (0.5,0.5) (using R recycling), and EM's final estimate was (0.361,0.639). Our initial guess for the $\mu_i$ was (55,80) (obtained by visually inspecting the unconditional estimated density plot), and EM's final guess was (54.6,80.1). That was pretty close to our initial guess, so we were lucky to have peaks to visually identify, thus aiding convergence. We see EM's estimates of the $\sigma_i$ above too.

Needless to say, there is more to be done. Is the 2-normal mixture a good model here? One might apply Kolmogorov-Smirnov as one approach to answering this (Section 8.6.)

## 13.3   Clustering

The goal is to try to find important subgroups in the data. Note that we do not know the groups beforehand; we are merely trying to find some. In machine learning terms, this is *unsupervised classification.*

Clearly this is related to the idea of mixture distributions, and indeed many methods formally or roughly assume a mixture of multivariate normal distributions. Since clustering is primarily used in two-variable settings, the assumption is then bivariate normal.

There is a vast literature on clustering, including many books, such as [23]. Many R packages are available as well; see a partial list in the CRAN Task View, Multivariate section.[2]

Let's discuss the oldest and simplest method, K-Means. It really has no assumptions, but it is presumed that the densities $f_{X|M}()$ are roughly mound-shaped.

The algorithm is iterative, of course. The user sets the number of clusters $k$, and makes initial guesses as to the centers of the clusters. Here is the

---

[2]https://cran.r-project.org/web/views/Multivariate.html

pseudocode:

```
set k and initial guesses for the k centers
do
    for i = 1,...,n
        find the cluster center j that case
            i is closest to
        assign case i to cluster j
    for each cluster m = 1,...,k
        find the new center of this cluster
until convergence
```

At any step, the center of a cluster is found by find the mean (x,y) coordinates of the data points currently in that cluster. Convergence is taken to be convergence of the cluster centers.

All this presumes the user knows the number of clusters beforehand. In simple algorithms such as this one, that is the case, and the user must experiment with several different values of $k$. The output of more sophisticated algorithms based on eigenvectors estimate the number of clusters along with estimating the centers.

The R function $kmeans()$ implements K-Means.

## 13.4 Exercises

**Mathematical problems:**

**1.** Say we have coins in a bag, with 80% balanced, probability 0.5 of heads, and 20% with heads probability 0.55. We choose a coin at random from the bag, then toss it five times. Using (13.21), find the variance of the number of heads we obtain.

**2.** In the Bonus Flips example, Section 13.1.3, find $Var(X)$.

**3.** In the Bus Ridership problem, Section 1.1, find $Var(L_2)$. (Suggestion: Condition on $L_1$.)

**4.** Let $N$ have a geometric distribution with parameter $p$. Find a closed-form expression for $E(N \mid N \leq k)$, $k = 1, 2, 3, \ldots$.

**Computational and data problems:**

**5.** Following up on the suggestion made at the end of Section 13.2.2.1, find the maximum discrepancy between the empirical cdf and the fitted mixture cdf, using Kolmogorov-Smirnov.

**6.** Another approach to assessing the fit in Section 13.2.2.1 would be to see if variances match well enough. Estimate $Var(X)$ in two ways:

- Find $s^2$ from the data.

- Estimate $Var(X)$ under the 2-normals model, using (13.21).

It would not be easy to go further — how close do the two variance estimates be in order to be a good fit? — but this could serve as an informal assessment.

# Chapter 14

# Multivariate Description and Dimension Reduction

*A model should be as simple as possible, but no simpler* — Albert Einstein

*If I had more time, I would have written a shorter letter* — attributed to various, e.g., Mark Twain[1]

Consider a dataset of $n$ cases, each of which has values of $p$ variables. We may for instance have a sample of $n = 100$ people, with data on $p = 3$ variables, height, weight and age.

In this era of Big Data, both $n$ and $p$ may be quite large, say millions of cases and thousands of variables. This chapter is primarily concerned with handling large values of $p$. Directly or indirectly, we will work on reducing $p$, termed *dimension reduction*.

- *Understanding the data:* We wish to have a more compact version of the data, with few variables. This facilitates "making sense" of the data: What is related to what? Are there interesting subgroups of the data, and so on?

- *Avoiding overfitting:* As you will see later, larger samples, i.e., large $n$, allows fitting more complex models, i.e., larger $p$. Problems occur

---

[1]Possibly the first known instance was by 17th century mathematician Blaise Pascal, "Je n'ai fait celle-ci plus longue que parce que je n'ai pas eu le loisir de la faire plus courte," roughly "I've made this one longer as I didn't have time to make it shorter."

when we fit a model that is more complex than our sample size $n$ can handle, termed *overfitting*.

There is also a third goal:

- *Managing the amount of computation:* Amazon has millions of users and millions of items. Think then of the ratings matrix, showing which rating each user gives to each item. It has millions of rows and millions of columns. Though the matrix is mostly empty — most users have not rated most items — it is huge, likely containing many terabytes of data. In order to manage computation, simplification of the data is key.

To be sure, in meeting these goals, we lose some information. But the gains may outweigh the losses.

In this chapter, we'll discuss two major methods for dimension reduction, one for continuous variables (Principal Components Analysis) and another for categorical ones (the log-linear model).

Note that **both methods will be presented as descriptive/exploratory in nature, rather than involving inference (confidence intervals and significance tests).** There are no commonly-used inference techniques for our first method, and though this is not true for our second method, it too will be presented as a descriptive/exploratory tool.

But first, let's take a closer look at overfitting.

# 14.1   What Is Overfitting Anyway?

## 14.1.1   "Desperate for Data"

Suppose we have the samples of men's and women's heights, $X_1, ..., X_n$ and $Y_1, ..., Y_n$. Assume for simplicity that the variance of height is the same for each gender, $\sigma^2$. The means of the two populations are designated by $\mu_1$ and $\mu_2$.

Say we wish to guess the height of a new person who we know to be a man but for whom we know nothing else. We do not see him, etc.

## 14.1.2 Known Distribution

Suppose for just a moment that we actually know the distribution of X, i.e., the *population* distribution of male heights. What would be the best constant $g'$ to use as our guess for a person about whom we know nothing other than gender?

Well, we might use mean squared error,

$$E[(g - X)^2] \tag{14.1}$$

as our criterion of goodness of guessing. But we already know what the best g is, from Section 4.24: The best g is $\mu_1$. Our best guess for this unseen man's height is the mean height of all men in the population, very intuitive.

## 14.1.3 Estimated Mean

Of course, we don't know $\mu_1$, but we can do the next-best thing, i.e., use an estimate of it from our sample.

The natural choice for that estimator would be

$$T_1 = \overline{X}, \tag{14.2}$$

the mean height of men in our sample.

But what if $n$ is really small, say $n = 5$? That's awfully small. We may wish to consider adding the women's heights to our estimate, in order to get a larger sample. Then we would estimate $\mu_1$ by

$$T_2 = \frac{\overline{X} + \overline{Y}}{2}, \tag{14.3}$$

It may at first seem obvious that $T_1$ is the better estimator. Women tend to be shorter, after all, so pooling the data from the two genders would induce a bias. On the other hand, we found in Section 4.24 that for any estimator,

$$\text{MSE} = \text{variance of the estimator} + \text{bias of the estimator}^2 \tag{14.4}$$

CHAPTER 14. DESCRIPTION AND DIMENSION

In other words, *some amount of bias may be tolerable*, if it will buy us a substantial reduction in variance. After all, women are not that much shorter than men, so the bias might not be too bad. Meanwhile, the pooled estimate should have lower variance, as it is based on $2n$ observations instead of $n$; (7.5) tells us that.

Before continuing, note first that $T_2$ is based on a simpler model than is $T_1$, as $T_2$ ignores gender. We thus refer to $T_1$ as being based on the more complex model.

Which one is better? The answer will need a criterion for goodness of estimation, which we will take to be mean squared error, MSE. So, the question becomes, which has the smaller MSE, $T_1$ or $T_2$? In other words:

Which is smaller, $E[(T_1 - \mu_1)^2]$ or $E[(T_2 - \mu_1)^2]$?

## 14.1.4 The Bias/Variance Tradeoff: Concrete Illustration

Let's find the biases of the two estimators.

- $T_1$

  From (7.4),

$$ET_1 = \mu_1 \tag{14.5}$$

  so

  bias of $T_1 = 0$

- $T_2$

$$
\begin{aligned}
E(T_2) &= E(0.5\overline{X} + 0.5\overline{Y}) \text{ (definition)} \tag{14.6}\\
&= 0.5E\overline{X} + 0.5E\overline{Y} \text{ (linearity of E())} \tag{14.7}\\
&= 0.5\mu_1 + 0.5\mu_2 \text{ [from (7.4)]} \tag{14.8}
\end{aligned}
$$

  So,

  bias of $T_2 = (0.5\mu_1 + 0.5\mu_2) - \mu_1 = 0.5(\mu_2 - \mu_1)$

On the other hand, $T_2$ has a smaller variance than $T_1$:

- $T_1$

  Recalling (7.5), we have

  $$Var(T_1) = \frac{\sigma^2}{n} \tag{14.9}$$

- $T_2$

$$
\begin{aligned}
Var(T_2) &= Var(0.5\overline{X} + 0.5\overline{Y}) & (14.10)\\
&= 0.5^2 Var(\overline{X}) + 0.5^2 Var(\overline{Y}) & (14.11)\\
&= 2 \cdot 0.25 \cdot \frac{\sigma^2}{n} \quad \text{[from 7.5]} & (14.12)\\
&= \frac{\sigma^2}{2n} & (14.13)
\end{aligned}
$$

These findings are highly instructive. You might at first think that "of course" $T_1$ would be the better predictor than $T_2$. But for a small sample size, the smaller (actually 0) bias of $T_1$ is not enough to counteract its larger variance. $T_2$ is biased, yes, but it is based on double the sample size and thus has half the variance.

So, under what circumstances will $T_1$ be better than $T_2$?

$$MSE(T_1) = \frac{\sigma^2}{n} + 0^2 = \frac{\sigma^2}{n} \tag{14.14}$$

$$MSE(T_2) = \frac{\sigma^2}{2n} + \left(\frac{\mu_1 + \mu_2}{2} - \mu_1\right)^2 = \frac{\sigma^2}{2n} + \left(\frac{\mu_2 - \mu_1}{2}\right)^2 \tag{14.15}$$

$T_1$ is a better predictor than $T_2$ if (14.14) is smaller than (14.15), which is true if

$$\left(\frac{\mu_2 - \mu_1}{2}\right)^2 > \frac{\sigma^2}{2n} \tag{14.16}$$

Granted, we don't know the values of the $\mu_1$ and $\sigma^2$. But the above analysis makes the point that under some circumstances, it really is better to pool the data in spite of bias.

## 14.1.5   Implications

You can see that $T_1$ is better only if either

- $n$ is large enough, or

- the difference in population mean heights between men and women is large enough, or

- there is not much variation within each population, e.g., most men have very similar heights

Since that third item, small within-population variance, is rarely seen, let's concentrate on the first two items. The big revelation here is that:

> A more complex model is more accurate than a simpler one only if either
>
> - we have enough data to support it, or
>
> - the complex model is sufficiently different from the simpler one

**In height/gender example above, if $n$ is too small, we are "desperate for data," and thus make use of the female data to augment our male data.** Though women tend to be shorter than men, the bias that results from that augmentation is offset by the reduction in estimator variance that we get. But if $n$ is large enough, the variance will be small in either model, so when we go to the more complex model, the advantage gained by reducing the bias will more than compensate for the increase in variance.

> **THIS IS AN ABSOLUTELY FUNDAMENTAL NO-TION IN STATISTICS/MACHINE LEARNING.** It will be key in Chapter 15 on predictive analytics.

This was a very simple example, but you can see that in complex settings, fitting too rich a model can result in very high MSEs for the estimates. In essence, everything becomes noise. (Some people have cleverly coined the term **noise mining**, a play on the term **data mining**.) This is the famous *overfitting* problem.

Note that of course (14.16) contains several unknown population quantities. We derived it here merely to establish a <u>principle</u>, namely that a more complex model may perform more poorly under some circumstances.

It would be possible, though, to make (14.16) into a practical decision tool, by estimating the unknown quantities, e.g., replacing $\mu_1$ by $\overline{X}$. This then creates possible problems with confidence intervals, whose derivation did not include this extra decision step. Such estimators, termed *adaptive*, are beyond the scope of this book.

# 14.2 Principal Components Analysis

Of the many methods for dimension reduction, one of the most commonly used is Principal Components Analysis, PCA. It is used both for dimension reduction and data understanding.

## 14.2.1 Intuition

Recall Figure 12.2. Let's call the two variables $X_1$ and $X_2$, with the corresponding axes in the graphs to be referred to as $t_1$ (sloping gently to the right) and $t_2$ (sloping steply upward and to the left). The graph was generated using simulated data, with a correlation of 0.8 between $X_1$ and $X_2$. Not surprisingly due to that high correlation, the "two-dimensional bell" is concentrated around a straight line, specifically the line

$$t_1 + t_2 = 1 \tag{14.17}$$

So there is high probability that

$$U_1 = X_1 + X_2 \approx 1 \tag{14.18}$$

i.e.,

$$X_2 \approx 1 - X_1 \tag{14.19}$$

In other words,

> To a large extent, there is only one variable here, $X_1$ (or other choices, e.g., $X_2$), not two.

Actually, the main variation of this data is along the line

$$t_1 + t_2 = 1 \tag{14.20}$$

The remaining variation is along the perpendicular line

$$t_1 - t_2 = 0 \tag{14.21}$$

Recall from Section 12.5 that the level sets in the three-dimensional bell are ellipses. The major and minor axes of the ellipse are (14.20) and (14.21). And the random variables $U_1 = X_1 - X_2$ and $U_2 = X_1 + X_2$ measure where we are along these two axes. Moreover, $X_1 + X_2$ and $X_1 - X_2$ are uncorrelated (Problem 1).

With that in mind, now suppose we have $p$ variables, $X_1, X_2, ..., X_p$, not just two. We can no longer visualize in higher dimensions, but as mentioned in (12.5), the level sets will be $p$-dimensional ellipsoids. These now have $p$ axes rather than just two, and we can define $p$ new variables, $Y_1, Y_2, ..., Y_p$ from the $X_i$, such that:

- The $Y_j$ are linear combinations of the $X_i$.

- The $Y_i$ are uncorrelated.

The $Y_j$ are called the *principal components* of the data.[2]

Now, what does this give us? The $Y_j$ carry the same information as the $X_i$ (since we have merely done a rotation of axes), but the benefit of using them accrues from ordering them by variance. We relabel the indices, taking $Y_1$ to be whichever original $Y_j$ as the largest variance, etc., so that

$$Var(Y_1) > Var(Y_2) > ... > Var(Y_p) \tag{14.22}$$

In our two-dimensional example above, recall that the data had most of its variation along the line (14.20), with much smaller remaining variation in the perpendicular direction (14.21). We thus saw that our data was largely one-dimensional.

---

[2]Recall that each $Y_j$ is some linear combination of the $X_i$. It is customary to not only refer to the $Y_j$ as the principal components, but also apply this term to those coefficient vectors.

Similarly, with $p$ variables, inspection of (14.22) may show that only a few of the $Y_j$, say $k$ of them, have substantial variances. We would then think of the data as essentially being $k$-dimensional, and use these $k$ variables in our subsequent analysis.

If so, we've largely or maybe completely abandoned using the $X_i$. Most of our future analysis of this data may be based on these new variables the $Y_1, ..., Y_k$.

Say we have $n$ cases in our data, say data on height, weight, age and income, denoted $X_1$, $X_2$, $X_3$ and $X_4$. We store all this in an $n \times 4$ matrix $Q$, with the first column being height, second being weight and so on. Suppose we decide to use the first two principal components, $Y_1$ and $Y_2$. So, $Y_1$ might be, say, 0.23 height + 1.16 weight + 0.12 age - 2.01 income. Then our new data is stored in an $n \times 2$ matrix $R$. So for instance Person 5 in our data will have his values of the $X_i$, i.e., height and so on, in row 5 of $Q$, while his values of the new variables will be in row 5 of $R$.

*Note the importance of the $Y_j$ being uncorrelated.* Remember, we are trying to find a subset of variables that are in some sense minimal. Thus we want no redundancy, and so the uncorrelated nature of the $Y_j$ is a welcome property.

## 14.2.2 Properties of PCA

As in the height/weight/age/income example above, but now being general, let $Q$ denote the original data matrix, with $R$ being the matrix of new data. Column $i$ of $Q$ is data on $X_i$, while column $j$ of $R$ is our data on $Y_j$. Say we use $k$ of the $p$ principal components, so $Q$ is $n \times p$ while $R$ is $n \times k$.

Let $U$ denote the matrix of eigenvectors of $A$, the covariance matrix of $Q$.[3] Then:

(a) $R = QU$.

(b) The columns of $U$ are orthogonal.

(c) The diagonal elements of $Cov(R)$, i.e., the variances of the principal components, are the eigenvalues of $A$, while the off-diagonal elements are all 0, i.e., the principal components are uncorrelated..

---

[3] From 11.38, that matrix is $Q'Q - \overline{Q}\,\overline{Q}'$, where $\overline{Q}$ is the $p \times 1$ vector of column averages of $Q$.

### 14.2.3   Example: Turkish Teaching Evaluations

The most commonly used R function for PCA is **prcomp()**. As with many R functions, it has many optional arguments; we'll take the default values here.

For our example, let's use the Turkish Teaching Evaluation data, available from the UC Irvine Machine Learning Data Repository [12]. It consists of 5820 student evaluations of university instructors. Each student evaluation consists of answers to 28 questions, each calling for a rating of 1-5, plus some other variables we won't consider here.

```
> turk <- read.csv('turkiye-student-evaluation.csv',
    header=TRUE)
> tpca <- prcomp(turk[,-(1:5)])
```

Let's explore the output. First, let's look at the standard deviations of the new variables, and the corresponding cumulative proportion of total variance in the data:

```
> tpca$sdev
 [1]  6.1294752 1.4366581 0.8169210 0.7663429
 [5]  0.6881709 0.6528149 0.5776757 0.5460676
 [9]  0.5270327 0.4827412 0.4776421 0.4714887
[13]  0.4449105 0.4364215 0.4327540 0.4236855
[17]  0.4182859 0.4053242 0.3937768 0.3895587
[21]  0.3707312 0.3674430 0.3618074 0.3527829
[25]  0.3379096 0.3312691 0.2979928 0.2888057
> tmp <- cumsum(tpca$sdev^2)
> tmp / tmp[28]
 [1]  0.8219815 0.8671382 0.8817389 0.8945877
 [5]  0.9049489 0.9142727 0.9215737 0.9280977
 [9]  0.9341747 0.9392732 0.9442646 0.9491282
[13]  0.9534589 0.9576259 0.9617232 0.9656506
[17]  0.9694785 0.9730729 0.9764653 0.9797855
[21]  0.9827925 0.9857464 0.9886104 0.9913333
[25]  0.9938314 0.9962324 0.9981752 1.0000000
```

This is striking, The first principal component (PC) already accounts for 82% of the total variance among all 28 questions. The first five PCs cover over 90%. This suggests that the designer of the evaluation survey could have written a much more concise survey instrument with almost the same utility.

The coefficients in the linear combinations that make up the principal components, i.e., our $U$ matrix above, are given in the columns of the **rotation** part of the object returned from **prcomp()**.

While we are here, let's check that the columns of $U$ are orthogonal, say the first two:

```
> t(tpca$rotation[,1]) %*% tpca$rotation[,2]
             [,1]
[1,] -2.012279e-16
```

Yes, 0 (about $-2 \times 10^{-16}$ with roundoff error).

And let's confirm that the off-diagonal elements are 0:

```
> r <- tpca$x
> cvr <- cov(r)
> max(abs(cvr[row(cvr) != col(cvr)]))
[1] 2.982173e-13
```

## 14.3   The Log-Linear Model

Suppose we have a dataset on physical characteristics of people, including variables on hair and eye color, and gender. These are *categorical* variables, alluding to the fact that they represent categories. Though we could use PCA to describe them (first forming indicator variables, e.g., brown, black, blue and so on for eyes), PCA may not do a good job here. An alternative is *log-linear models*, which model various types of interactions among a group of categorical variables, ranging from full independence to different levels of partial independence.

In terms of our two goals set at the opening of this chapter, this method is more often used for understanding, but can be quite helpful in terms of avoiding *avoding overfitting*.

This is a very rich area of methodology, about which many books have been written [9]. We mere introduce the topic here.

### 14.3.1   Example: Hair Color, Eye Color and Gender

As a motivating example, consider the dataset **HairEyeColor**, built in to R.

The reader can view the dataset simply by typing **HairEyeColor** at the R prompt. It's arranged as an R **table** type. Online help is available via **?HairEyeColor**. The variables are: Hair, denoted below by $X^{(1)}$; Eye, $X^{(2)}$; and Sex, $X^{(3)}$.

Let $X_r^{(s)}$ denote the value of $X^{(s)}$ for the $r^{th}$ person in our sample, $r = 1, 2, ..., n$. Our data are the counts

$$N_{ijk} = \text{number of r such that } X_r^{(1)} = i, X_r^{(2)} = j \text{ and } X_r^{(3)} = k \quad (14.23)$$

**Overview of the data:**

```
> HairEyeColor
, , Sex = Male

        Eye
Hair    Brown Blue Hazel Green
  Black    32   11    10     3
  Brown    53   50    25    15
  Red      10   10     7     7
  Blond     3   30     5     8


, , Sex = Female

        Eye
Hair    Brown Blue Hazel Green
  Black    36    9     5     2
  Brown    66   34    29    14
  Red      16    7     7     7
  Blond     4   64     5     8
```

Note that this is a three-dimensional array, with Hair being rows, Eye being columns, and Sex being layers. The data above show, for instance, that there are 25 men with brown hair and hazel eyes, i.e., $N_{231} = 25$.

Let's check this:

```
> HairEyeColor[2,3,1]
[1] 25
```

Here we have a three-dimensional **contingency table**. Each $N_{ijk}$ value is a **cell** in the table. If we have $k$ categorical variables, the table is said to be $k$-dimensional.

## 14.3.2 Dimension of Our Data

At the outset of this chapter, we discussed dimension reduction in terms of number of variables. But in this case, a better definition would be number of parameters. Here the latter are the cell probabilities; $p_{ijk}$ be the population probability of a randomly-chosen person falling into cell $ijk$, i.e.,

$$p_{ijk} = P\left(X^{(1)} = i \text{ and } X^{(2)} = j \text{ and } X^{(3)} = k\right) = E(N_{ijk})/n \quad (14.24)$$

There are $4 \times 4 \times 2 = 32$ of them. In fact, there really are only 31, as the remaining one is equal to 1.0 minus the sum of the others.

Reducing dimension here will involve simpler models of this data, as we will see below.

## 14.3.3 Estimating the Parameters

As mentioned, the $p_{ijk}$ are population parameters. How can we estimate them from our sample data, the $N_{ijk}$?

Recall the two famous methods for parameter estimation in Chapter 8, the Method of Moments (MM) and Maximum Likelihood Estimation (MLE). Without any further assumptions on the data, MM and MLE just yield the "natural" estimator,

$$\widehat{p}_{ijk} = N_{ijk}/n \quad (14.25)$$

But things change when we add further assumptions. We might assume, say, that hair color, eye color and gender are independent, i.e.,

$$P\left(X^{(1)} = i \text{ and } X^{(2)} = j \text{ and } X^{(3)} = k\right) \quad (14.26)$$

$$= P\left(X^{(1)} = i\right) \cdot P\left(X^{(2)} = j\right) \cdot P\left(X^{(3)} = k\right)$$

Under this assumption, the number of parameters is much lower. There are 4 for $P(X^{(1)} = i)$, 4 for $P(X^{(2)} = j)$, and 2 for $P(X^{(2)} = k)$. But just as in the assumptionless case we have 31 parameters instead of 32, the numbers 4, 4 and 2 above are actually 3, 3 and 1. In other words, we have only 7 parameters if we assume independence, rather than 31. This is dimension reduction!

Though these 7 parameters might be estimated by either MM or MLE, the machinery that has been built up uses MLE.

Models of partial independence — in between the 31-parameter and 7-parameter models — are also possble. For instance, hair color, eye color and gender may not be fully independent, but maybe hair and eye color are independent within each gender. In other words, hair and eye color are conditionally independent, given gender. Thus one task for the analyst is to decide which of the several possible models best fits the data.

Among others, R's **loglin()** function fits log-linear models (so called because the models fit the quantities $\log(p_{ijk})$). An introduction is given in Section 14.5.2.


## 14.4   Mathematical Complements

### 14.4.1   Statistical Derivation of PCA

One can derive PCA via (B.15), but are more statistical approach is more instructive, as follows.

Let $A$ denote the sample covariance matrix of our $X$ data (Section 11.7). As before, let $X_1, ..., X_p$ be the variables in our original data, say height, weight and age. In our data, write $B_{ij}$ for the value of $X_j$ in data point $i$ of our data.

Let $U$ denote some linear combination of our $X_i$,

$$U = d_1 X_1 + ... + d_p X_p = d'(X_1, ..., X_p)' \qquad (14.27)$$

From (11.37), we have

$$Var(U) = d'Ad \qquad (14.28)$$

Recall that the first principal component has maximal variance, so we wish to maximize $Var(U)$. But of course, that variance would have no limit without a constraint. so we require $d = (d_1, ..., d_p)'$ to have length 1,

$$d'd = 1 \qquad (14.29)$$

Thus we ask, What value of $d$ will maximize (14.28), subject to (14.29)?

In math, to maximize a function $f(t)$ subject to a constraint $g(t) = 0$, we use the method of *Lagrange multipliers*. This involves introducing a new "artificial" variable $\lambda$, and maximizing

$$f(t) + \lambda g(t) \qquad (14.30)$$

with respect to both $t$ and $\lambda$.[4] Here we set $f(d)$ to (14.28), and $g(d)$ to $d'd - 1$, i.e., we will maximize

$$d'Ad + \lambda(d'd - 1) \qquad (14.31)$$

From (B.20), we have

$$\frac{\partial}{\partial d} d'Ad = 2A'd = 2Ad \qquad (14.32)$$

(using the fact that $A$ is a symmetric matrix). Also,

$$\frac{\partial}{\partial d} d'd = 2d \qquad (14.33)$$

So, differentiating (14.31) with respect to $d$, we have

$$0 = 2Ad + \lambda 2d \qquad (14.34)$$

so

$$Ad = -\lambda d \qquad (14.35)$$

In other words, the coefficient vector of the first principal component $d$ needs to be an eigenvalue of $A$! In fact, one can show that the coefficient vectors of *all* of the principal components of our data must be eigenvectors of $A$.

---

[4]Maximizing with respect to the latter is just a formality, a mechanism to force $g(t) = 0$,

## 14.5  Computational Complements

### 14.5.1  R Tables

Say we have two variables, the first having levels 1 and 2, and the second having levels 1, 2 and 3. Suppose our data frame **d** is

```
> d
  V1 V2
1  1  3
2  2  3
3  2  2
4  1  1
5  1  2
```

The first person (or other entity) in our dataset has $X^{(1)} = 1$, $X^{(2)} = 3$, then $X^{(1)} = 2$, $X^{(2)} = 3$ in the second, and so on. The function **table()** does what its name implies: It tabulates the counts in each cell:

```
> table(d)
    V2
V1   1 2 3
   1 1 1 1
   2 0 1 1
```

This says there was one instance in which $X^{(1)} = 1, X^{(2)} = 3$ etc., but no instances of $X^{(1)} = 2, X^{(2)} = 1$.

### 14.5.2  Some Details on Log-Linear Models

These models can be quite involved, and the reader is urged to consult one of the many excellent books on the topic, e.g., [9]. We'll give a brief example here, again using the **HairEyeColor** dataset.

Consider first the model that assumes full independence:

$$p_{ijk} = P\left(X^{(1)} = i \text{ and } X^{(2)} = j \text{ and } X^{(3)} = k\right) \quad (14.36)$$

$$= P\left(X^{(1)} = i\right) \cdot P\left(X^{(2)} = j\right) \cdot P\left(X^{(3)} = k\right) \quad (14.37)$$

Taking logs of both sides in (14.37), we see that independence of the three variables is equivalent to saying

$$\log(p_{ijk}) = a_i + b_j + c_k \tag{14.38}$$

for some numbers $a_i$, $b_j$ and $c_j$; e.g.,

$$b_2 = \log[P(X^{(2)} = 2)] \tag{14.39}$$

So, independence corresponds to additivity on the log scale. On the other hand, if we assume instead that Sex is independent of Hair and Eye but that Hair and Eye are not independent of each other, our model would include an $i, j$ interaction term, as follows.

We would have

$$p_{ijk} \;\; = \;\; P\left(X^{(1)} = i \text{ and } X^{(2)} = j\right) \cdot P\left(X^{(3)} = k\right) \tag{14.40}$$

so we would set

$$\log(p_{ijk}) = a_{ij} + b_k \tag{14.41}$$

Most formal models rewrite the first term as

$$a_{ij} = u + v_i + w_j + r_{ij} \tag{14.42}$$

Here we have written $P\left(X^{(1)} = i \text{ and } X^{(2)} = j\right)$ as a sum of an "overall effect" u, "main effects" $v_i$ and $w_j$, and "interaction effects."[5]

### 14.5.2.1   Parameter Estimation

Remember, whenever we have parametric models, the statistician's "Swiss army knife" is Maximum Likelihood Estimation (MLE, Section 8.4.3). That is what is most often used in the case of log-linear models.

---

[5]There are also constraints, taking the form that various sums must be 0. These follow naturally from the formulation of the model, but are beyond the scope of this book.

How, then, do we compute the likelihood of our data, the $N_{ijk}$? It's actually quite straightforward, because the $N_{ijk}$ have a *multinomial distribution*, a generalization of the binomial distribution family. (The latter assumes two categories, while multinomial accommodates multiple categories.) Then the likelihood function is

$$L = \frac{n!}{\Pi_{i,j,k} N_{ijk}!} p_{ijk}^{N_{ijk}} \qquad\qquad (14.43)$$

We then write the $p_{ijk}$ in terms of our model parameters.

$$p_{ijk} = \exp(u + v_i + w_j + r_{ik} + b_k) \qquad\qquad (14.44)$$

We then substitute (14.44) in (14.43), and maximize the latter with respect to the $u, v_i,...$, subject to constraints as mentioned earlier.

The maximization may be messy. But certain cases have been worked out in closed form, and in any case today one would typically do the computation by computer. In R, for example, there is the **loglin()** function for this purpose, illustrated below.

### 14.5.2.2   The loglin() Function

Continue to consider the HairEyeColor dataset.

We'll use the built-in R function **loglin()**, whose input data must be of class "**table**".

Let's fit a model (as noted, for the $N_{ijk}$ rather than the $p_{ijk}$) in which hair and eye color are independent of gender, but not with each other, i.e., the model (14.42):

```
fm <- loglin(HairEyeColor, list(c(1, 2),3),
    param=TRUE,fit=TRUE)
```

Our model is input via the argument **margin**, here **list(c(1, 2),3)**. It's an R list of vectors specifying the model. For instance **c(1,3)** specifies an interaction between variables 1 and 3, and **c(1,2,3)** means a three-way interaction. Once a higher-order interaction is specified, we need not specify its lower-order "subset." If, say, we specify **c(2,5,6)**, we need not specify **c(2,6)**.

### 14.5.2.3 Informal Assessment of Fit

Let's again consider the case of brown-haired, hazel-eyed men, i.e.,

$$p_{231} = \frac{EN_{231}}{n} \tag{14.45}$$

The model fit, for $EN_{231}$ is

```
> fm$fit[2,3,1]
[1] 25.44932
```

This compares to the actual observed value of 25 we saw earlier. Let's see all of the fitted values:

```
> fm$fit
, , Sex = Male

       Eye
Hair      Brown      Blue     Hazel     Green
  Black 32.047297  9.425676  7.069257  2.356419
  Brown 56.082770 39.587838 25.449324 13.667230
  Red   12.253378  8.011824  6.597973  6.597973
  Blond  3.298986 44.300676  4.712838  7.540541

, , Sex = Female

       Eye
Hair      Brown      Blue     Hazel     Green
  Black 35.952703 10.574324  7.930743  2.643581
  Brown 62.917230 44.412162 28.550676 15.332770
  Red   13.746622  8.988176  7.402027  7.402027
  Blond  3.701014 49.699324  5.287162  8.459459
```

Here are the observed values again:

```
> HairEyeColor
, , Sex = Male

       Eye
Hair    Brown Blue Hazel Green
  Black    32   11    10     3
  Brown    53   50    25    15
  Red      10   10     7     7
```

```
Blond       3    30      5       8

, , Sex = Female

         Eye
Hair    Brown Blue Hazel Green
   Black    36    9      5       2
   Brown    66   34     29      14
   Red      16    7      7       7
   Blond     4   64      5       8
```

Actually, the values are not too far off. Informally, we might say that conditional independence of gender from hair/eye color describes the data well. (And makes intuitive sense.)

The "elephant in the room" here is sampling variation. As stated at the outset of this chapter, we are presenting the log-linear model merely as a descriptive tool, just as PCA is used for decription rather than inference. But we must recognize that with another sample from the same population, the fitted values might be somewhat different. Indeed, what appears to be a "nice fit" may actually be overfitting.

Standard errors would be nice to have. Unfortunately, most books and software packages for the log-linear model put almost all of their focus on significance testing, rather than confidence intervals, and standard errors are not available.

There is a solution, though, the *Poisson trick* As noted, the $N_{ijk}$ have a multinomial distribution. Imagine for a moment, though, that they are independent, Poisson-distributed random variables. Then $n$, the total cell count, is now random, so call it $N$. The point is that, conditionally on $N$, the $N_{ijk}$ have a multinomial distribution. We can then use software for the Poisson model, e.g., one of the options in R's **glm()**, to analyze the multinomial case, including computation of standard errors. See [31] for details.

## 14.6   Exercises

**Mathematical problems:**

1. In the discussion following (14.21), Show that $X_1 + X_2$ and $X_1 - X_2$ are uncorrelated.

**Computational and data problems:**

**2**. Load R's built-in **UCBAdmissions** dataset, discussed in Exercise 6, Chapter 10. Fit a log-linear model in which Sex is independent of Admission and Department, while the latter two are not independent of each other.

**3**. Download the YearPredictionMSD dataset from the UCI Machine Learning Repository. It contains various audio meausurements, lots of them, and thus is good place to try PCA for dimension reduction. Explore this: Note: This is a very large dataset. If you exceed the memory of your computer, you may wish to try the **bigstatsr** package.

# Chapter 15

# Predictive Modeling

*Prediction is hard, especially about the future* — baseball legend Yogi Berra

Here we are interested in relations between variables. Specifically:

> In *regression analysis*, we are interested in the relation of one variable, $Y$, with one or more others, which we will collectively write as the vector $X$. Our tool is the conditional mean, $E(Y \mid X)$.

Note carefully that *many types of methods that go by another name are actually regression methods*. Examples are the *classification problem* and *machine learning*, which we will see are special cases of regression analysis.

Note too that though many users of such methods association the term *regression function* with a linear model, the actual, more much general meaning is that of a conditional mean of one variable given one or more others. In the special case in which the variable to be predicted is an indicator (Sec. 4.4) variable, the conditional mean becomes the conditional probability of a 1 value.

## 15.1  Example: Heritage Health Prize

A company named Kaggle (kaggle.com) has an interesting business model — they host data science contests, with cash prizes. One of the more lucrative ones was the Heritage Health Prize [22]:

> More than 71 million individuals in the United States are admitted to hospitals each year, according to the latest survey from the American Hospital Association. Studies have concluded that in 2006 well over $30 billion was spent on unnecessary hospital admissions. Is there a better way? Can we identify earlier those most at risk and ensure they get the treatment they need? The Heritage Provider Network (HPN) believes that the answer is "yes".

Here $Y$ was 1 for hospitalization, 0 if not, and $X$ consisted of various pieces of information about a member's health history. The prize for the best predictive model was $500,000.

## 15.2   The Goals: Prediction and Description

Before beginning, it is important to understand the typical goals in regression analysis.

- **Prediction:** Here we are trying to predict one variable from one or more others, as with the Heritage Health contest above.

- **Description:** Here we wish to determine which of several variables have a greater effect on a given variable, and whether the effect is positive or negative.[1] An important special case is that in which we are interested in determining the effect of one predictor variable, **after the effects of the other predictors are removed.**

### 15.2.1   Terminology

We will term the components of the vector $X = (X^{(1)}, ..., X^{(r)})'$ as *predictor variables*, alluding to the Prediction goal. They are also called *explanatory variables*, highlighting the Description goal. In the machine learning realm, they are called *features*.

The variable to be predicted, $Y$, is often called the *response variable*, or the *dependent variable*. Note that one or more of the variables — whether the

---

[1]One must be careful not to attribute causation. The word "effect" here only refers to relations.

predictors or the response variable — may be indicator variables (Section 4.4). For instance, a gender variable may be coded 1 for male, 0 for female. Another name for predictor variables of that type is *dummy variables*, and it will be seen later in this chapter that they play a major role.

Methodology for this kind of setting is called *regression analysis*. If the response variable $Y$ is an indicator variable, as with the Kaggle example above, we call this the *classification problem*. The classes here are Hospitalize and Not Hospitalize. In many applications, there are more than two classes, in which case $Y$ will be a vector of indicator variables.

# 15.3 What Does "Relationship" Mean?

Suppose we are interested in exploring the relationship between adult human height $H$ and weight $W$.

As usual, we must first ask, *what does that really mean?* What do we mean by "relationship"? Clearly, there is no exact relationship; for instance, a person's weight is not an exact function of his/her height.

Effective use of the methods to be presented here requires an understanding of what exactly is meant by the term *relationship* in this context.

## 15.3.1 Precise Definition

Intuitively, we would guess that mean weight increases with height. To state this precisely, the key word in the previous sentence is *mean*.

Define

$$m_{W;H}(t) = E(W \mid H = t) \tag{15.1}$$

This looks abstract, but it is just common-sense stuff. Consider $m_{W;H}(68)$, for instance; that would be the mean weight of all people in the subpopulation of height 68 inches. By contrast, $EW$ is the mean weight of all people in the population as a whole.

The value of $m_{W;H}(t)$ varies with $t$, and we would expect that a graph of it would show an increasing trend in $t$, reflecting that taller people tend to be heavier. Note again the phrase *tend to*; it's not true for individuals, but is true for *mean* weight as a function of height.

We call $m_{W;H}$ the *regression function of W on H*. In general, $m_{Y;X}(t)$ means the mean of $Y$ among all units in the population for which $X = t$.[2] Note the word *population* in that last sentence. The function $m()$ is a population function. **The issue will be how to estimate it from sample data.**

So we have:

> **Major Point 1:** When we talk about the *relationship* of one variable to one or more others, we are referring to the regression function, which expresses the population mean of the first variable as a function of the others. Again: The key word here is *mean!*

As noted, in real applications, we don't know $E(Y \mid X)$, and must estimate it from sample data. How can we do this? Toward that end, let's suppose we have a random sample of 1000 people from city of Davis, with

$$(H_1, W_1), ..., (H_{1000}, W_{1000}) \tag{15.2}$$

being their heights and weights. As in previous sample data we've worked with, we wish to use this data to estimate population values. But the difference here is that we are estimating a whole function now, the entire curve $m_{W;H}(t)$ as $t$ varies. That means we are estimating infinitely many values, with one $m_{W;H}(t)$ value for each $t$.[3] How do we do this?

One approach would be as follows. Say we wish to find $\widehat{m}_{W;H}(t)$ (note the hat, for "estimate of"!) at $t = 70.2$. In other words, we wish to estimate the mean weight — in the population — among all people of height 70.2 inches. What we could do is look at all the people in our sample who are within, say, 1.0 inch of 70.2, and calculate the average of all their weights. This would then be our $\widehat{m}_{W;H}(t)$.

---

[2]The word "regression" is an allusion to the famous comment of Sir Francis Galton in the late 1800s regarding "regression toward the mean." This referred to the fact that tall parents tend to have children who are less tall — closer to the mean — with a similar statement for short parents. The predictor variable here might be, say, the father's height $F$, with the response variable being, say, the son's height $S$. Galton was saying that $E(S \mid F \text{ tall}) < F$.

[3]Of course, the population of Davis is finite, but there is the conceptual population of all people who could live in Davis, past, present and future.

## 15.3.2 Parametric Models for the Regression Function m()

Recall that in Chapter 8, we compared fitting parametric and nonparametric ("model-free") models to our data. For instance, we were interested in estimating the population density of the BMI data. On the one hand, we could simply plot a histogram. On the other hand, we could assume a gamma distribution model, and estimate the gamma density by estimating the two gamma parameters, using MM or MLE.

Note that there is a bit of a connection with our current regression case, in that estimation of a density $f()$ again involves estimating infinitely many parameters — the values at infinitely many different $t$. If the gamma is a good model, we need estimate only two parameters, quite a savings from infinity!

In the regression case, we again can choose from nonparametric and parameters models. The approach described above — averaging the weights of all sample people with heights within 1.0 inch of 68 — is nonparametric. There are many nonparametric methods like this, and in fact most of today's machine learning methods are variants of this. But the traditional method is to choose a parametric model for the regression function. That way we estimate only a finite number of quantities instead of an infinite number.

Typically the parametric model chosen is linear, i.e., we assume that $m_{W;H}(t)$ is a linear function of $t$:

$$m_{W;H}(t) = ct + d \tag{15.3}$$

for some constants $c$ and $d$. If this assumption is reasonable — meaning that though it may not be exactly true it is reasonably close — then it is a huge gain for us over a nonparametric model. Do you see why? Again, the answer is that instead of having to estimate an infinite number of quantities, we now must estimate only two quantities — the parameters $c$ and $d$.

Equation (15.3) is thus called a *parametric* model of $m_{W;H}()$. The set of straight lines indexed by $c$ and $d$ is a two-parameter family, analogous to parametric families of distributions, such as the two-parametric gamma family; the difference, of course, is that in the gamma case we were modeling a density function, and here we are modeling a regression function.

Note that $c$ and $d$ are indeed population parameters in the same sense that, for instance, $r$ and $\lambda$ are parameters in the gamma distribution family. We must estimate $c$ and $d$ from our sample data, which we will address shortly.

So we have:

> **Major Point 2:** The function $m_{W;H}(t)$ is a population entity, so we must estimate it from our sample data. To do this, we have a choice of either assuming that $m_{W;H}(t)$ takes on some parametric form, or making no such assumption.
>
> If we opt for a parametric approach, the most common model is linear, i.e., (15.3). Again, the quantities $c$ and $d$ in (15.3) are population values, and as such, we must estimate them from the data.

## 15.4  Estimation in Linear Parametric Regression Models

So, how can we estimate these population values $c$ and $d$? We'll go into details in Section 15.9, but here is a preview:

Recall the "Useful Fact" on page 71: The minimum expected squared error guess for a random variable is its mean. This implies that the best estimator of a random variable in a *conditional* distribution is the *conditional* mean. Taken together with the principle of iterated expectation, (13.2) and (13.10), we have this, say for the human weight and height example:

The minimum value of the quantity

$$E\left[(W - g(H))^2\right] \qquad (15.4)$$

over all possible functions $g(H)$, is attained by setting

$$g(H) = m_{W;H}(H) \qquad (15.5)$$

In other words, $m_{W;H}(H)$ is the optimal predictor of $W$ among all possible functions of H, in the sense of minimizing mean squared prediction error.[4]

Since we are assuming the model (15.3), this in turn means that:

---

[4]But if we wish to minimize the mean absolute prediction error, $E\left(|W - g(H)|\right)$, the best function turns out to be is $g(H) = \text{median}(W|H)$.

The quantity

$$E\left[(W - (uH + v))^2\right] \qquad (15.6)$$

is minimized by setting $u = c$ and $v = d$.

This then gives us a clue as to how to estimate $c$ and $d$ from our data, as follows.

If you recall, in earlier chapters we've often chosen estimators by using sample analogs, e.g., $s^2$ as an estimator of $\sigma^2$. Well, the sample analog of (15.6) is

$$\frac{1}{n} \sum_{i=1}^{n} [W_i - (uH_i + v)]^2 \qquad (15.7)$$

Here (15.6) is the mean squared prediction error using $u$ and $v$ in the population, and (15.7) is the mean squared prediction error using $u$ and $v$ in our sample. Since $u = c$ and $v = d$ minimize (15.6), it is natural to estimate $c$ and $d$ by the $u$ and $v$ that minimize (15.7).

Using the "hat" notation common for estimators, we'll denote the $u$ and $v$ that minimize (15.7) by $\hat{c}$ and $\hat{d}$, respectively. These numbers are then the classical *least-squares estimators* of the population values $c$ and $d$.

> **Major Point 3:** In statistical regression analysis, one often uses a linear model as in (15.3), estimating the coefficients by minimizing (15.7).

We will elaborate on this in Section 15.9.

## 15.5  Example: Baseball Data

This data on 1015 major league baseball players was obtained courtesy of the UCLA Statistics Department, and is included in my R package **freqparcoord** [30].

Let's do a regression analysis of weight against height.

## 15.5.1   R Code

First, we load the data, and take a look at the first few records:

```
> library(freqparcoord)
> data(mlb)
> head(mlb)
          Name Team        Position Height
1    Adam_Donachie  BAL       Catcher     74
2       Paul_Bako  BAL       Catcher     74
3 Ramon_Hernandez  BAL       Catcher     72
4    Kevin_Millar  BAL First_Baseman     72
5     Chris_Gomez  BAL First_Baseman     73
6   Brian_Roberts  BAL Second_Baseman     69
  Weight   Age PosCategory
1    180 22.99     Catcher
2    215 34.69     Catcher
3    210 30.78     Catcher
4    210 35.43   Infielder
5    188 35.71   Infielder
6    176 29.39   Infielder
```

Now run R's **lm()** ("linear model") function to perform the regression analysis:

```
> lm(Weight ~ Height,data=mlb)

Call:
lm(formula = Weight ~ Height, data = mlb)

Coefficients:
(Intercept)          Height
   -151.133           4.783
```

We can get a little more information by calling **summary()**:

```
> lmout <- lm(Weight ~ Height,data=mlb)
> summary(lmout)
...
Coefficients:
              Estimate Std. Error t value Pr(>|t|)
(Intercept) -151.1333    17.6568   -8.56   <2e-16
Height         4.7833     0.2395   19.97   <2e-16
```

```
(Intercept)  ***
Height       ***
. . .
Multiple R-squared:   0.2825,      Adjusted R-squared:
0.2818
. . .
```

(Not all the output is shown here, as indicated by the "...")

The function **summary()** is an example of a *generic* function in R, covered earlier in Section 8.9.1. There are some other R issues here, but we'll relegate them to the Computational Complements section at the end of the chapter.

In our call to **lm()**, we are asking R to use the **mlb** data frame, regressing the **Weight** column against the **Height** column.

Next, note that **lm()** returns a lot of information (even more than shown above), all packed into an object of type "lm".[5] By calling **summary()** on that object, I obtained some of the information. First, we see that the sample estimates of $c$ and $d$ are

$$\hat{d} = -155.092 \qquad\qquad (15.8)$$

$$\hat{c} = 4.841 \qquad\qquad (15.9)$$

In other words, our estimate for the function giving mean weight in terms of height is

$$\text{mean weight} = \text{-155.092} + 4.841 \text{ height} \qquad (15.10)$$

Do keep in mind that this is just an estimate, based on the sample data; it is not the population mean-weight-versus-height function. So for example, our *sample estimate* is that an extra inch in height corresponds on average to about 4.8 more pounds in weight. The exact population value will probably be somewhat different.

We can form a confidence interval to make that point clear, and get an idea of how accurate our estimate is. The R output tells us that the standard error of $\hat{c}$ is 0.240. Making use of (10.6,) we add and subtract 1.96 times

---

[5] R class names are quoted.

this number to $\widehat{c}$ to get our interval: (4.351,5.331). So, we are about 95% confident that the true slope, $c$, is contained in that interval.

Note the column of output labled "t value." As discussed in Section 10.7, the t-distribution is almost identical to N(0,1) except for small samples. We are testing

$$H_0 : c = 0 \qquad\qquad (15.11)$$

The p-value given in the output, less than $2 \times 10^{-16}$, is essentially that obtained from (10.11): The latter has value 19.97, the output tells us, and the area under the N(0,1) density to the right of that value turns out to be under $10^{-16}$.

It is customary in statistics to place one, two or three asterisks next to a p-value, depending on whether the latter is less than 0.05, 0.01 or 0.001, respectively. R generally follows this tradition, as seen here.

Finally, the output shown above mentions an $R^2$ value. What is this? To answer this question, let's consider the predicted values we would get from (15.10). If there were a player for whom we knew only height, say 72.6 inches, we would guess his weight to be

$$-155.092 + 4.841(72.6) = 190.3646 \qquad\qquad (15.12)$$

Fine, but what about a player *in* our dataset, whose weight we *do* know? Why guess it? The point here is to assess how good our prediction model is, by predicting players with known weights, and the comparing our predictions with the true weights. We saw above, for instance, that the first player in our dataset had height 74 and weight 180. Our equation would predict his weight to be 203.142, a rather substantial error.

**Definition 29** *Suppose a model (parametric or nonparametric) for the regression function of Y on X is fit to sample data $(Y_i, X_i)$, $i = 1, ..., n$ (with X possibly vector-valued). Let $\widehat{Y}_i$ denote the fitted value in case i. $R^2$ is the squared sample correlation between the predicted values, i.e., the $\widehat{Y}_i$ and the actual $Y_i$.*

The $R^2$ here, 0.2825, is modest. Height has some value in predicting weight, but it is limited. One sometimes hears statements of the form, "Height explains 28% of the variation in weight," providing further interpretation for the figure.

Finally, recall the term *bias* from Chapter 7. It turns out that $R^2$ is biased upward, i.e., it tends to be higher than the true population value. There is a quantity *adjusted* $R^2$, reported by **summary()**, that is designed to be less biased.

Here the adjusted version differs only slightly from the ordinary $R^2$. If there is a substantial difference, it is an indication of overfitting. This was discussed in Chapter 14, and will be returned to later.

## 15.6 Multiple Regression

Note that in the regression expression $E(Y \mid X = t)$, $X$ and $t$ could be vector-valued. For instance, we could have $Y$ be weight and have $X$ be the pair

$$X = \left(X^{(1)}, X^{(2)}\right)' = (H, A)' = \text{(height, age)}' \qquad (15.13)$$

so as to study the relationship of weight with height and age. If we used a linear model, we would write for $t = (t_1, t_2)'$,

$$m_{W;H,A}(t) = \beta_0 + \beta_1 t_1 + \beta_2 t_2 \qquad (15.14)$$

In other words

$$\text{mean weight} = \beta_0 + \beta_1 \text{ height} + \beta_2 \text{ age} \qquad (15.15)$$

Once again, keep in mind that (15.14) and (15.15) are models for the population. We assume that (15.14), (15.15) or whichever model we use is an exact representation of the relation in the population. And of course, our derivations below assume our model is correct.

(It is traditional to use the Greek letter $\beta$ to name the coefficient vector in a linear regression model.)

So for instance $m_{W;H,A}(68, 37.2)$ would be the mean weight in the population of all people having height 68 and age 37.2.

In analogy with (15.7), we would estimate the $\beta_i$ by minimizing

$$\frac{1}{n}\sum_{i=1}^{n}\left[W_i - (u + vH_i + wA_i)\right]^2 \qquad (15.16)$$

with respect to $u$, $v$ and $w$. The minimizing values would be denoted $\widehat{\beta}_0$, $\widehat{\beta}_1$ and $\widehat{\beta}_2$.

We might consider adding a third predictor, gender:

$$\text{mean weight} = \beta_0 + \beta_1 \text{ height} + \beta_2 \text{ age} + \beta_3 \text{ gender} \qquad (15.17)$$

where **gender** is an indicator variable, 1 for male, 0 for female. Note that we would not have two gender variables, one for each gender, since knowledge of the value of one such variable would tell us what the other one is. (It would also make a certain matrix noninvertible, as we'll discuss later.)

## 15.7   Example: Baseball Data (cont'd.)

So, let's regress weight against height and age:

```
> summary(lm(Weight ~ Height+Age,data=mlb))
...
Coefficients:
              Estimate Std. Error t value Pr(>|t|)
(Intercept) -187.6382    17.9447  -10.46  < 2e-16
Height         4.9236     0.2344   21.00  < 2e-16
Age            0.9115     0.1257    7.25 8.25e-13

(Intercept) ***
Height      ***
Age         ***
...
Multiple R-squared:  0.318,      Adjusted R-squared:
0.3166
```

So, our regression function coefficient estimates are $\hat{\beta}_0 = -187.6382$, $\hat{\beta}_1 = 4.9236$ and $\hat{\beta}_2 = 0.9115$. Note that the $R^2$ values increased.

This is an example of the Description goal in some regression applications. We might be interested in whether players gain weight as they age. Many people do, of course, but since athletes try to keep fit, the answer is less clear for them. That is a Description question, not Prediction.

Here we estimate from our sample data that 10 years' extra age results, on average, of a weight gain about about 9.1 pounds — for people of a given height. This last condition is very important.

## 15.8   Interaction Terms

Equation (15.14) implicitly says that, for instance, the effect of age on weight is the same at all height levels. In other words, the difference in mean weight between 30-year-olds and 40-year-olds is the same regardless of whether we are looking at tall people or short people. To see that, just plug 40 and 30 for age in (15.14), with the same number for height in both, and subtract; you get $10\beta_2$, an expression that has no height term.

That assumption is not a good one, since the weight gain in aging tends to be larger for tall people than for short ones. If we don't like this assumption, we can add an *interaction term* to (15.14), consisting of the product of the two original predictors. Our new predictor variable $X^{(3)}$ is equal to $X^{(1)}X^{(2)}$, and thus our regression function is

$$m_{W;H}(t) = \beta_0 + \beta_1 t_1 + \beta_2 t_2 + \beta_3 t_1 t_2 \qquad (15.18)$$

If you perform the same subtraction described above, you'll see that this more complex model does not assume, as the old did, that the difference in mean weight between 30-year-olds and 40-year-olds is the same regardless of whether we are looking at tall people or short people.

Though the idea of adding interaction terms to a regression model is tempting, it can easily get out of hand. If we have $k$ basic predictor variables, then there are $\binom{k}{2}$ potential two-way interaction terms, $\binom{k}{3}$ three-way terms and so on. Unless we have a very large amount of data, we run a big risk of overfitting (Chapter 16). And with so many interaction terms, the model would be difficult to interpret.

We can add even more interaction terms by introducing powers of variables, say the square of height in addition to height. Then (15.18) would become

$$m_{W;H}(t) = \beta_0 + \beta_1 t_1 + \beta_2 t_2 + \beta_3 t_1 t_2 + \beta_4 t_1^2 \qquad (15.19)$$

This square is essentially the "interaction" of height with itself. If we believe the relation between mean weight and height is quadratic, this might be worthwhile, but again, this means more and more predictors.

So, we may have a decision to make here, as to whether to introduce interaction terms. For that matter, it may be the case that age is actually not that important, so we even might consider dropping that variable altogether.

## 15.9   Parametric Estimation

So, how did R compute those estimated regression coefficients? Let's take a look.

### 15.9.1   Meaning of "Linear"

Hey, wait a minute...how could we call that quadratic model (15.19) "linear"? Here's why:

Here we model $m_{Y;X}$ as a linear function of $X^{(1)}, ..., X^{(r)}$:

$$m_{Y;X}(t) = \beta_0 + \beta_1 t^{(1)} + ... + \beta_r t^{(r)} \qquad (15.20)$$

A key point is that the term **linear regression** does NOT necessarily mean that the graph of the regression function is a straight line or a plane. Instead, the word *linear* refers to the regression function being linear in the parameters. So, for instance, (15.19) is a linear model; if for example we multiple each of $\beta_0, \beta_1, ..., \beta_4$ by 8, then $m_{W;H}(t)$ is multiplied by 8.

A more literal look at the meaning of "linear" comes from the matrix formulation (15.28) below.

### 15.9.2   Random-X and Fixed-X Regression

Consider our earlier example of estimating the regression function of weight on height. To make things, simple, say we sample only 5 people, so our data is $(H_1, W_1), ..., (H_5, W_5)$. and we measure height to the nearest inch.

In our "notebook" view, each line of our notebook would have 5 heights and 5 weights. Since we would have a different set of 5 people on each line, the $H_1$ column will generally have different values from line to line, though occasionally two consecutive lines will have the same value. $H_1$ is a random variable. We call regression analysis in this setting *random-X* regression.

We could, on the other hand, set up our sampling plan so that we sample one person each of heights 65, 67, 69, 71 and 73. These values would then stay the same from line to line. The $H_1$ column, for instance, would consist entirely of 65s. This is called *fixed-X regression*.

So, the probabilistic structure of the two settings is different. However, it turns out not to matter much, for the following reason.

Recall that the definition of the regression function, concerns the *conditional* distribution of $W$ given $H$. So, our analysis below will revolve around that conditional distribution, in which case $H$ becomes nonrandom anyway.

## 15.9.3 Point Estimates and Matrix Formulation

So, how do we estimate the $\beta_i$? Keep in mind that the $\beta_i$ are population values, which we need to estimate from our data. How do we do that? For instance, how did R compute the $\widehat{\beta_i}$ in Section 15.5? As previewed above, usual method is least-squares. Here we will go into the details.

For concreteness, think of the baseball data, and let $H_i$, $A_i$ and $W_i$ denote the height, age and weight of the $i^{th}$ player in our sample, i = 1,2,...,1015. As mentioned, the estimation methodology involves finding the values of $u_i$ that minimize the sum of squared differences between the actual $W$ values and their predicted values using the $u_i$:

$$\sum_{i=1}^{1015} [W_i - (u_0 + u_1 H_i + u_2 A_i)]^2 \quad (15.21)$$

When we find the minimizing $u_i$, we will set our estimates for the population regression coefficients $\beta_i$ in (15.20):

$$\widehat{\beta_0} = u_0 \quad (15.22)$$

$$\widehat{\beta_1} = u_1 \quad (15.23)$$

$$\widehat{\beta_2} = u_2 \tag{15.24}$$

Obviously, this is a calculus problem. We set the partial derivatives of (15.21) with respect to the $u_i$ to 0, giving us three linear equations in three unknowns, and then solve.

However...everything becomes easier if we write all this in linear algebra terms. Define

$$V = \begin{pmatrix} W_1 \\ W_2 \\ \dots \\ W_{1015} \end{pmatrix}, \tag{15.25}$$

$$u = \begin{pmatrix} u_0 \\ u_1 \\ u_2 \end{pmatrix} \tag{15.26}$$

and

$$Q = \begin{pmatrix} 1 & H_1 & A_1 \\ 1 & H_2 & A_2 \\ \dots \\ 1 & H_{1015} & A_{1015} \end{pmatrix} \tag{15.27}$$

Then

$$E(V \mid Q) = Q\beta \tag{15.28}$$

To see this, look at the first player, of height 74 and age 22.99 (Section 15.5.1). We are modeling the mean weight in the population for all players of that height and weight as

$$\text{mean weight} = \beta_0 + \beta_1 \, 74 + \beta_2 \, 22.99 \tag{15.29}$$

The top row of $Q$ will be $(1,74,22.99)$, so the top row of $Q\beta$ will be $\beta_0 + \beta_1 \, 74 + \beta_2 \, 22.99$ — which exactly matches (15.29). Note the need for the 1s column in $Q$, in order to pick up the $\beta_0$ term.

We can write (15.21) as

$$(V - Qu)'(V - Qu) \tag{15.30}$$

(Again, just look at the top row of $V - Qu$ to see this.)

Whatever vector $u$ minimizes (15.30), we set our estimated $\beta$ vector $\widehat{\beta} = (\widehat{\beta_0}, \widehat{\beta_1}, \widehat{\beta_2})'$ to that $u$.

As noted, we know that we can minimize (15.30) by taking the partial derivatives with respect to $u_0, u_1, ..., u_r$, setting them to 0 and so on. But there is a matrix formulation here too. It is shown in Section B.5.1 that, the solution is

$$\hat{\beta} = (Q'Q)^{-1}Q'V \tag{15.31}$$

For the general case (15.20) with $n$ observations (n = 1015 in the baseball data), the matrix $Q$ has $n$ rows and $r + 1$ columns. Column $i + 1$ has the sample data on predictor variable $i$.

Note that we are conditioning on $Q$ in (15.28). This is the standard approach, especially since that is the case of nonrandom $X$. Thus we will later get conditional confidence intervals, which is fine. To avoid clutter, we will not show the conditioning explicitly, and thus for instance will write, for example, $Cov(V)$ instead of $Cov(V|Q)$.

It turns out that $\hat{\beta}$ is an unbiased estimate of $\beta$:[6]

$$
\begin{aligned}
E\hat{\beta} &= E[(Q'Q)^{-1}Q'V] \quad \text{(15.31)} & (15.32)\\
&= (Q'Q)^{-1}Q'EV \quad \text{(linearity of E())} & (15.33)\\
&= (Q'Q)^{-1}Q' \cdot Q\beta \quad \text{(15.28)} & (15.34)\\
&= \beta & (15.35)
\end{aligned}
$$

In some applications, we assume there is no constant term $\beta_0$ in (15.20). This means that our $Q$ matrix no longer has the column of 1s on the left end, but everything else above is valid.[7]

---

[6]Note that here we are taking the expected value of a vector, as in Chapter 11.

[7]In the **lm()** call, we must write -1, e.g., lm(Weight $\sim$Height-1,data=mlb)

## 15.9.4    Approximate Confidence Intervals

As seen above, **lm()** gives you standard errors for the estimated coefficients. Where do they come from? And what assumptions are needed?

As usual, we should not be satisfied with just point estimates, in this case the $\widehat{\beta}_i$. We need an indication of how accurate they are, so we need confidence intervals. In other words, we need to use the $\widehat{\beta}_i$ to form confidence intervals for the $\beta_i$.

For instance, recall that our **lm()** analysis of the baseball players indicated that they do gain weight as they age, about a pound per year. The goal there would primarily be Description, specifically assessing the impact of age. That impact is measured by $\beta_2$. Thus, we want to find a confidence interval for $\beta_2$.

Equation (15.31) shows that the $\widehat{\beta}_i$ are linear combinations — hence sums! — of the components of $V$, i.e., the $W_j$ in the weight-height-age example. So, the Central Limit Theorem implies that the $\widehat{\beta}_i$ are approximately normally distributed.[8] That in turn means that, in order to form confidence intervals, we need standard errors for the $\widehat{\beta}_i$. How will we get them? (Or, equivalenty, where does **lm()** get them for its output?)

A typical assumption made in regression books is that the distribution of $Y$ given $X$ is normal. In regressing weight on height, for instance, this would mean that in any fixed height group, say 68.3 inches, the population distribution of weight is normal. We will NOT make this assumption, and as pointed out above, the CLT is good enough for us to get our confidence intervals. Note too that the so-called "exact" Student-t intervals are illusory, since no distribution in real life is exactly normal.

However, we do need to add an assumption in order to get standard errors:

$$Var(Y|X=t) = \sigma^2 \qquad (15.36)$$

for all $t$. Note that this and the independence of the sample observations (e.g., the various people sampled in the weight/height example are independent of each other) implies that

$$Cov(V|Q) = \sigma^2 I \qquad (15.37)$$

---

[8]The form of the CLT presented in this book is for sums of independent, identically distributed random variables. But there are versions for the independent but not identically distributed case [25].

where $I$ is the usual identiy matrix (1s on the diagonal, 0s off diagonal).

Be sure you understand what this means. In the weight/height example, for instance, it means that the variance of weight among 72-inch tall people is the same as that for 65-inch-tall people. That is not quite true — the taller group has larger variance — but it is a standard assumption that we will make use of here.

(A better solution to this problem is available in the *sandwich estimator*, which is implemented for example in the **car** package [16].)

We can derive the covariance matrix of $\hat{\beta}$ as follows. To avoid clutter, let $B = (Q'Q)^{-1}$. A theorem from linear algebra says that $Q'Q$ is symmetric and thus $B$ is too. So $B' = B$, a point to be used below. Another theorem says that for any conformable matrices $U$ and $V$, then $(UV)' = V'U'$. Armed with that knowledge, here we go:

$$
\begin{aligned}
Cov(\hat{\beta}) &= Cov(BQ'V) \ ((15.31)) & (15.38) \\
&= BQ'Cov(V)(BQ')' \ (11.35) & (15.39) \\
&= BQ'\sigma^2 I(BQ')' \ (15.37) & (15.40) \\
&= \sigma^2 BQ'QB \ \text{(lin. alg.)} & (15.41) \\
&= \sigma^2(Q'Q)^{-1} \ \text{(def. of B)} & (15.42)
\end{aligned}
$$

Whew! That's a lot of work for you, if your linear algebra is rusty. But it's worth it, because (15.42) now gives us what we need for confidence intervals. Here's how:

First, we need to estimate $\sigma^2$. Recall that for any random variable $U$, $Var(U) = E[(U - EU)^2]$, we have

$$
\begin{aligned}
\sigma^2 &= Var(Y|X = t) & (15.43) \\
&= Var(Y|X^{(1)} = t_1, ..., X^{(r)} = t_r) & (15.44) \\
&= E\left[\{Y - m_{Y;X}(t)\}^2\right] & (15.45) \\
&= E\left[(Y - \beta_0 - \beta_1 t_1 - ... - \beta_r t_r)^2\right] & (15.46)
\end{aligned}
$$

Thus, a natural estimate for $\sigma^2$ would be the sample analog, where we replace $E()$ by averaging over our sample, and replace population quantities by sample estimates:

$$s^2 = \frac{1}{n} \sum_{i=1}^{n} (Y_i - \hat{\beta}_0 - \hat{\beta}_1 X_i^{(1)} - \ldots - \hat{\beta}_r X_i^{(r)})^2 \qquad (15.47)$$

As in Chapter 7, this estimate of $\sigma^2$ is biased, and classically one divides by $n - (r + 1)$ instead of $n$. But again, it's not an issue unless $r + 1$ is a substantial fraction of $n$, in which case you are overfitting and shouldn't be using a model with so large a value of $r$ anyway.

So, the estimated covariance matrix for $\hat{\beta}$ is

$$\widehat{Cov}(\hat{\beta}) = s^2 (Q'Q)^{-1} \qquad (15.48)$$

The diagonal elements here are the squared standard errors (recall that the standard error of an estimator is its estimated standard deviation) of the $\hat{\beta}_i$. (And the off-diagonal elements are the estimated covariances between the $\hat{\beta}_i$.)

## 15.10    Example: Baseball Data (cont'd.)

Let us use the generic function **vcov()** to obtain the estimated covariance matrix of the vector $\hat{\beta}$ for our baseball data.

```
> lmout <- lm(Weight ~ Height + Age,data=mlb)
> vcov(lmout)
               (Intercept)          Height              Age
(Intercept) 322.0112213   -4.119253943   -0.633017113
Height        -4.1192539    0.054952432    0.002432329
Age           -0.6330171    0.002432329    0.015806536
```

For instance, the estimated variance of $\hat{\beta}_1$ is 0.054952432. Taking the square root, we see that the standard error of $\hat{\beta}_1$ 0.2344. This matches our earlier call to **summary(lm())**, which of course got its number from the same source.

But now we can find more. Say we wish to compute a confidence interval for the population mean weight of players who are 72 inches tall and age

30. That quantity is equal to

$$\beta_0 + 72\beta_1 + 30\beta_2 = (1, 72, 30)\beta \qquad (15.49)$$

which we will estimate by

$$(1, 72, 30)\widehat{\beta} \qquad (15.50)$$

Thus, using (11.37), we have

$$\widehat{Var}(\widehat{\beta}_0 + 72\widehat{\beta}_1 + 30\widehat{\beta}_2) = (1, 72, 30)A \begin{pmatrix} 1 \\ 72 \\ 30 \end{pmatrix} \qquad (15.51)$$

where A is the matrix in the R output above.

The square root of this quantity is the standard error of $\widehat{\beta}_0 + 72\widehat{\beta}_1 + 30\widehat{\beta}_2$. We add and subtract 1.96 times that square root to $\widehat{\beta}_0 + 72\widehat{\beta}_1 + 30\widehat{\beta}_2$, and then have an approximate 95% confidence interval for the population mean weight of players who are 72 inches tall and age 30.

# 15.11   Dummy Variables

Many datasets include *categorical* or *nominal* variables, with these terms indicating that such a variable codes categories or names of categories.

Consider a study of software engineer productivity [18]. The authors of the study predicted $Y$ = number of person-months needed to complete a project, from $X^{(1)}$ = size of the project as measured in lines of code, $X^{(2)}$ = 1 or 0 depending on whether an object-oriented or procedural approach was used, and other variables.

$X^{(2)}$ is an indicator variable, often called a "dummy" variable in the regression context. Let's generalize that a bit. Suppose we are comparing two different object-oriented languages, C++ and Java, as well as the procedural language C. Then we could change the definition of $X^{(2)}$ to have the value 1 for C++ and 0 for non-C++, and we could add another variable, $X^{(3)}$, which has the value 1 for Java and 0 for non-Java. Use of the C language would be implied by the situation $X^{(2)} = X^{(3)} = 0$.

Say in the original coding of our dataset, there had been a single variable Language, coded 0, 1 or 2 for C++, Java and C, respectively. There are several important points here.

- We do NOT want to represent Language by a single value having the values 0, 1 and 2, which would imply that C has, for instance, double the impact of Java.

- We would thus convert the single variable Language to the two dummy variables, $X^{(2)}$ and $X^{(3)}$.

- As mentioned, we would NOT create three dummies. It's clear that two suffice, and use of C is implied by $X^{(2)} = X^{(3)} = 0$. In fact, having three would cause $Q'Q$ to be noninvertible in (15.31). (Exercise 2, end of this chapter.)

- When R reads a dataset into a data frame, it will notice that a variable is categorical, and enter it into the data frame as a **factor**, which is an R data type specifically for categorical variables. If you subsequently call **lm()**, the latter will automatically convert factors to dummies, taking care to make only $k - 1$ dummies from a factor with $k$ levels.

## 15.12    Classification

In prediction problems, in the special case in which $Y$ is an indicator variable, with the value 1 if the object is in a class and 0 if not, the regression problem is called the *classification problem*.

We'll formalize this idea in Section 15.12.1, but first, here are some examples:

- Is a patient likely to develop diabetes? This problem has been studied by many researchers, e.g., [40]. We have already seen the Pima data in Chapter 7, where the predictors were number of times pregnant, plasma glucose concentration, diastolic blood pressure, triceps skin fold thickness, serum insulin level, body mass index, diabetes pedigree function and age.

- Is a disk drive likely to fail soon? This has been studied for example in [34]. $Y$ was 1 or 0, depending on whether the drive failed, and the predictors were temperature, number of read errors, and so on.

- An online service has many customers come and go. It would like to predict who is about to leave, so as to offer them a special deal for staying with this firm [45].

- Of course, a big application is computer vision [26].

In all of the above examples but the last, there are just two classes, e.g., diabetic or nondiabetic in the first example. In the last example, there are usually many classes. If we are trying to recognize handwritten digits 0-9, for instance, there are 10 classes. With facial recognition, the number of classes could be in the millions or more.

## 15.12.1    Classification = Regression

All of the many machine learning algorithms, despite their complexity, really boil down to regression at their core. Here's why:

As we have frequently noted, the mean of any indicator random variable is the probability that the variable is equal to 1 (Section 4.4). Thus in the case in which our response variable $Y$ takes on only the values 0 and 1, i.e., classification problems, the regression function reduces to

$$m_{Y;X}(t) = P(Y = 1 | X = t) \tag{15.52}$$

(Remember that $X$ and $t$ are typically vector-valued.)

As a simple but handy example, suppose $Y$ is gender (1 for male, 0 for female), $X^{(1)}$ is height and $X^{(2)}$ is weight, i.e., we are predicting a person's gender from the person's height and weight. Then for example, $m_{Y;X}(70, 150)$ is the probability that a person of height 70 inches and weight 150 pounds is a man. Note again that this probability is a population quantity, the fraction of men among all people of height 70 and weight 150 in our population.

One can prove rather easily that:

> Given $X = t$, the optimal prediction rule, i.e., the one minimizing the overall population misclassification rate, is to predict that $Y = 1$ if and only if $m_{Y;X}(t) > 0.5$.

So, if we known a certain person is of height 70 and weight 150, our best guess for the person's gender is to predict the person is male if and only if $m_{Y;X}(70, 150) > 0.5$.

## 15.12.2    Logistic Regression

Remember, we often try a parametric model for our regression function first, as it means we are estimating a finite number of quantities, instead of an infinite number.[9] Probably the most commonly-used classcation model is that of the *logistic function* (often called "logit"). Its $r$-predictor form is

$$m_{Y;X}(t) = P(Y = 1 | X = t) = \frac{1}{1 + e^{-(\beta_0 + \beta_1 t_1 + \ldots + \beta_r t_r)}} \qquad (15.53)$$

### 15.12.2.1    The Logistic Model: Motivations

The logistic function itself,

$$\ell(u) = \frac{1}{1 + e^{-u}} \qquad (15.54)$$

has values between 0 and 1, and is thus a good candidate for modeling a probability. Also, it is monotonic in $u$, making it further attractive, as in many classification problems we believe that $m_{Y;X}(t)$ should be monotonic in the predictor variables.

But there are additional reasons to use the logit model, as it includes many common parametric models for $X$. To see this, note that we can write, for vector-valued discrete $X$ and $t$,

$$
\begin{aligned}
P(Y = 1 | X = t) &= \frac{P(Y = 1 \text{ and } X = t)}{P(X = t)} \\
&= \frac{P(Y = 1)P(X = t | Y = 1)}{P(X = t)} \\
&= \frac{qP(X = t | Y = 1)}{qP(X = t | Y = 1) + (1 - q)P(X = t | Y = 0)} \\
&= \frac{1}{1 + \frac{(1-q)P(X=t|Y=0)}{qP(X=t|Y=1)}}
\end{aligned}
$$

---

[9]A nonparametric approach would be something like the following. Consider predicting gender fom height and weight as in the example above. We could find all the people in our sample data of height and weight near 70 and 150, respectively, and then compute the proportion of people in that set who are male. This would be our estimated probability of male for the given height and weight.

where $q = P(Y = 1)$ is the proportion of members of the population that have $Y = 1$. Keep in mind that this probability is unconditional!

If $X$ is a continuous random vector, then the analog is

$$P(Y = 1 | X = t) = \frac{1}{1 + \frac{(1-q)f_{X|Y=0}(t)}{qf_{X|Y=1}(t)}} \qquad (15.55)$$

Now for simplicity, suppose $X$ is scalar, i.e., $r = 1$. And suppose that, given $Y$, $X$ has a normal distribution. In other words, within each class, $Y$ is normally distributed. Suppose also that the two within-class variances of $X$ are equal, with common value $\sigma^2$, but with means $\mu_0$ and $\mu_1$. Then

$$f_{X|Y=i}(t) = \frac{1}{\sqrt{2\pi}\sigma} \exp\left[-0.5\left(\frac{t - \mu_i}{\sigma}\right)^2\right] \qquad (15.56)$$

After doing some elementary but rather tedious algebra, (15.55) reduces to the logistic form

$$\frac{1}{1 + e^{-(\beta_0 + \beta_1 t)}} \qquad (15.57)$$

where

$$\beta_0 = -\ln\left(\frac{1-q}{q}\right) + \frac{\mu_0^2 - \mu_1^2}{2\sigma^2}, \qquad (15.58)$$

and

$$\beta_1 = \frac{\mu_1 - \mu_0}{\sigma^2}, \qquad (15.59)$$

**In other words, if $X$ is normally distributed in both classes, with the same variance but different means, then $m_{Y;X}()$ has the logistic form!** And the same is true if $X$ is multivariate normal in each class, with different mean vectors but equal covariance matrices. (The algebra is even more tedious here, but it does work out.) Given the central importance of the multivariate normal family — the word *central* here is a pun, alluding to the (multivariate) Central Limit Theorem — this makes the logit model even more useful.

### 15.12.2.2    Estimation and Inference for Logit

We fit a logit model in R using the **glm()** function, with the argument **family=binomial**. The function finds Maximum Likelihood Estimates (Section 8.4.3) of the $\beta_i$.[10]

The output gives standard errors for the $\widehat{\beta}_i$ as in the linear model case. This enables the formation of confidence intervals and significance tests on individual $\widehat{\beta}_i$. For inference on linear combinations of the $\widehat{\beta}_i$, use the **vcov()** function as in the linear model case.

## 15.12.3    Example: Forest Cover Data

Let's look again at the forest cover data we saw in Section 10.6.1. Recall that this application has the Prediction goal, rather than the Description goal;[11] We wish to predict the type of forest cover. There were seven classes of forest cover.

## 15.12.4    R Code

For simplicity, let's restrict analysis to classes 1 and 2, so we have a two-class problem.[12] Create a new variable to serve as $Y$, recoding the 1,2 class names to 1,0:

```
> cvr1 <- cvr[cvr[,55] <= 2,]
> dim(cvr1)  # most cases still there
[1] 495141    55
> cvr1[,55] <- as.integer(cvr1[,55] == 1)
```

Let's see how well we can predict a site's class from the variable HS12 (hillside shade at noon, named V8 in the data) that we investigated earlier, now using a logistic model. (Of course, a better analysis would use more predictors.)

```
> g <- glm(V55 ~ V8,data=cvr1,family=binomial)
```

The result was:

---

[10]As in the case of linear regression, estimation and inference are done conditionally on the values of the predictor variables $X_i$.

[11]Recall these concepts from Section 15.2.

[12]This will be generalized in Section 15.12.5.1.

```
> summary(g)
...
Coefficients:
              Estimate Std. Error z value
(Intercept)  0.9647878  0.0351373   27.46
V8          -0.0055949  0.0001561  -35.83
             Pr(>|z|)
(Intercept)  <2e-16 ***
V8           <2e-16 ***
...
Number of Fisher Scoring iterations: 4
```

## 15.12.5 Analysis of the Results

You'll immediately notice the similarity to the output of **lm()**.[13] In particular, note the Coefficients section. There we have the estimates of the population coefficients $\beta_i$, their standard errors, and p-values for the tests of $H_0 : \beta_i = 0$.

We see that for example $\widehat{\beta}_1 = -0.01$. This is tiny, reflecting our analysis of this data in Chapter 10. There we found that the estimated mean values of HS12 for cover types 1 and 2 were 223.4 and 225.3, a difference of only 1.9, minuscule in comparison to the estimated means themselves. That difference in essence now gets multiplied by 0.01. Let's see the effect on the regression function, i.e., the probability of cover type 1 given HS12. In other words, let's imagine two forest sites, with unknown cover type, but known HS12 values 223.8 and 226.8 that are right in the center of the HS12 distribution for the two cover types. What would we predict for the cover types to be for those two sites?

Plugging in to (15.53), the results are 0.328 and 0.322, respectively. Remember, these numbers are the estimated probabilities that we have cover type 1, given HS12. So, our guess — predicting whether we have cover type 1 or 2 — isn't being helped much by knowing HS12; the probabilities of cover type 1 are very close to each other (and we would guess No in each case).

In other words, HS12 isn't having much effect on the probability of cover

---

[13]Did you notice that the last column is labled "z value" rather than "t-value" as before? The latter came from a Student t-distribution, which assumed that the distribution of $Y$ given $X$ was exactly normal. As we have discussed, that assumption is usually unrealistic, so we relied on the Central Limit Theorem. For larger $n$ is doesn't matter much anyway. Here, though, there is no exact test, so even R is resorting to the CLT.

type 1, and so it cannot be a good predictor of cover type.

**And yet...** the R output says that $\beta_1$ is "significantly" different from 0, with a tiny p-value of $2 \times 10^{-16}$. Thus, we see once again that significance testing does not achieve our goal.

### 15.12.5.1  Multiclass Case

So far, we have just restricted to two classes, cover types 1 and 2. How do we handle the problem of seven classes?

One approach to this problem would be to run seven logit models. The first would predict type 1 vs. others, the second would predict type 2 vs. others, then type 3 vs. others and so on. Given a new case for prediction, we would find the seven estimated probabilities of $Y = 1$ (with $Y$ playing a diff role in each one); we would then guess the cover type to be the one that turns out to have the highest probability.

To learn more about the multiclass case, see [29].

## 15.13    Machine Learning:  Neural Networks

Though typically presented in terms worthy of science fiction — the name *machine learning* itself sounding SciFi-ish — the fact is that machine learning (ML) techniques are simply nonparametric regression methods. Here we will discuss the ML method that generally gets the most attention, neural networks (NNs). Various other techniques are also popular, such as *random forests, boosting* and *Support Vector Machines.*[14]

### 15.13.1    Example:  Predicting Vertebral Abnormalities

Here six predictors are used to guess one of three vertebral conditions, normal (NO), disk hernia (DH) and spondylolisthesis (SL). Figure 15.1 was generated by the R NN package **neuralnet**.

[14]Technically, our treatment here will cover just *feedforward* NNs. Another type, *convolutional* NNs, is popular in image classification applications. These really are not any different from basic NNs; the "C" part consists of standard image-processing operations that long predate NNs and are widely used in non-NN contexts. *Recurrent* NNs, popular for text classication, allow loop connections in the network but otherwise have standard NN structure.

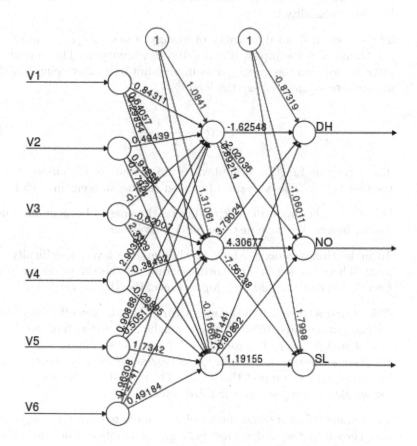

Error: 43.000304   Steps: 1292

Figure 15.1: Vertebrae data

There are three *layers* here, vertical columns of circles. The data flow is from left to right. The values of the six predictors for a particular patient, *V1* through *V6*, enter on the left, and our predicted class for that patient comes out of the rightmost layer. (Actually, the three outputs are class probabilities, and our predicted class will be taken to be the one with highest probability.)

The circles are called *neurons* or simply *units*. At each unit, a linear combination of the inputs is computed as an output. The outputs of one layer are fed into the next layer as inputs, but only after being run through an *activation function*, say the logit[15]

$$a(t) = \frac{1}{1 + e^{-t}} \qquad (15.60)$$

The activation function is applied to the outputs of the circles (except for the first layer). The activation function can be different in each layer.

Other than the input and output layers, the ones in between are known as *hidden layers*. We have just one hidden layer here.

To make things concrete, look at that middle layer, specifically the top unit. There are six inputs, corresponding to the six predictor variables. Each input has a *weight*. *V1*, for instance, has the weight 0.84311.

This at first sounds like the action of **lm()**, but the key difference is that we are computing three different sets of weights, one for each of the three units in that middle layer. The weight of *V1* into the second unit is 0.64057. To allow for a constant term as in linear regression models, there are also "1" inputs, seen at the top of the figure. The outputs of the second layer also get weights, for input into the third layer.

The number of layers, and the number of units per layer, are *hyperparameters*, chosen by the analyst, just as, e.g., the number of bins in a histogram is set by the analyst.

So, how are the weights determined? It's an iterative process (note the word "steps" in the caption of the figure), in which we are trying to minimize some loss function, typically total squared prediction error as in (15.30). The process can become quite complex, and in fact much of NN technology is devoted to making the iteration process faster and more accurate; indeed, in some cases, it may be difficult even to get the process to converge. We

---

[15]This of course is the function used in logistic regression, but there really is no connection. In each case, one needs an increasing function with values in (0,1), to produce probabilities, conditions that this function satisfies.

will not pursue that here, other than to note that many of the hyperparameters in any NN implementation are devoted to improving in those senses.

## 15.13.2   But What Is *Really* Going On?

With all this complexity, it is easy to miss how NNs work. To gain insight, consider an activation function $a(t) = t^2$. This is not a common one at all, but let's start with it.

As noted, the inputs to the second layer are linear combinations of the first layer values. But due to the activation function, the outputs of that second layer will be squared, so that the inputs of the third layer are — read carefully — linear combinations of *squares* of linear combinations of *V1* through *V6*. That means quadratic polynomials in *V1* through *V6*. If we were to have several more hidden layers, then the next layer would output polynomials of degree 4, then degree 8 and so on.

What about other activation functions $a(t)$? For any polynomial $a(t)$, you can see again that we will have polynomials of higher and higher degree as we go from layer to layer.

And what about (15.60)? Recall from calculus that one can approximate a function like that with a Taylor series — i.e., a polynomial! And even some common activation functions like one called *ReLU* that lack Taylor series can still be approximated by polynomials.[16]

In other words:

> NN models are closely related to polynomial regression.

## 15.13.3   R Packages

There are many R packages available for NNs. At present, the most sophisticated is **keras**, an R implementation of a popular general algorithm of the same name. The **kerasformula** package, acting as a wrapper to **keras**, gives the user a more "R-like" interface.[17] Here is a typical call pattern:

```
units <- c(5,2,NA)
layers <- list(units=units,
    activation=c('relu','relu','linear'))
kfout <- kms(y ~ .,data=z,layers=layers)
```

---

[16]$reLU(t) = max(0, t)$

[17]For installation tips, see https://keras.rstudio.com.

In the first line, we are specifying hidden layers consisting of 5 and 2 units, followed by a layer that simply passes through output from the previous layers. In the second line, we are specifying activation functions, reLU for the first two layers, with 'linear' again meaning passing data straight through. This is because we have a regression application here. For a classification problem, we would probably specify 'softmax' for our last activation function, which outputs the index of the largest input.

In general NN applications, regardless of implementation, it is recommended that typically one should *center and scale* one's data, as this has been found to help convergence properties.[18] One might, for instance, apply R's **scale()** function to the predictor data, which subtracts the means and divides by the standrd deviations. Note that this also means that in predicting new cases in the future, the same scaling must be applied to them. See the R online help on **scale()** to see how to save the original scaling values and then apply the same values later.

One might scale the response variable too. In **kerasformula**, that variable is scaled down to [0,1], by subtracting the lower bound and dividing by the range. For example, say the response values range from 1 to 5. Then we subtract 1 and divide by 5 - 1 = 4. Since the predicted values come back on this same scale, we must multiply by 4 and add 1 in order to return to the original scale.

# 15.14    Computational Complements

## 15.14.1    Computational Details in Section 15.5.1

Consider the line

```
> lm(Weight ~ Height,data=mlb)
```

in Section 15.5.1. It looks innocuous, but there is more than meets the eye here.

As seen by the '>' symbol, we executed this R's interactive mode. In that mode, any expresion we type will be printed out. The call to **lm()** returns an object, so it is printed out. What exactly is happening there? The object, say **o**, is an instance of the S3 class **"lm"**, actually quite complex. So, what does it mean to "print" a complicated object?

---

[18]One common issue is the "broken clock problem" in which the algorithm converges but all the predicted values are identical!

The answer lies in the fact that **print()** is an R generic function. The R interpreter, when asked to print **o**, will *dispatch* the print to the print function tailored to the "**lm**" class, **print.lm()**. The author of that function had to decide what kinds of information should be printed out.

Similarly, in

```
> summary(lmout)
```

the call was dispatched to **summary.lm()**, and the author of that function needed to decide what information to print out.

### 15.14.2    More Regarding glm()

R's **glm()** function is a generalization of **lm()**, and in fact "glm" stands for Generalized Linear Model. Here's why:

Consider (15.53). Though it is clearly nonlinear in the parameters $\beta_i$, one immediately notices that there is a linear form inside the expression, specifically in the exponent of $e$. In fact, the reader should verify that in Sec. (15.12.2.1),

$$w(u) = -\ln\left(\frac{1 - \ell(u)}{\ell(u)}\right) = u \qquad (15.61)$$

In other words, even though the logit itself is not linear in $u$, a function of the logit *is* linear in $u$. In GLM terms, this is called the *link function*; it *links* the regression function with a linear form in the $\beta_i$ [14].

There are various other GLM models, such as *Poisson regression*. There the conditional distribution of $Y$ given $X = t$ is assumed to be Poisson with $\lambda = \beta_0 + \beta_1 t_1 + ... + \beta_r t_r$. There the link function is simply **log()**.

As we saw, the link function is specified in **glm()** via the argument **family**, e.g., **family = binomial** for the logit link. Note that **family** must be a function, and in fact there are functions **binomial**, **poisson** and so on that compute the various link functions.

## 15.15   Exercises

**Mathematical problems:**

**1**. Consider the marbles example, Section 11.1.1. Note that these probabilities are the distribution, i.e., population values. Find the value of the population regression function $m_{Y;B}(j)$, $j = 0, 1, 2, 3$.

**2**. In the C++/Java/C example in Section 15.11, it was stated that we should not have three dummy variables to represent Language, it would render $Q'Q$ noninvertible. Prove this. Hint: Consider the vector sum of the three columns in $Q$ corresponding to these languages.)

**3**. Suppose $(X, Y)$, with both components scalars, has a bivariate normal distribution, with mean vector $\mu$ and covariance matrix $\Sigma$. Show that $m_{Y;X}(t) = \beta_0 + \beta_1 t$, i.e., the linear model holds, and find the $\beta_i$ in terms of $\mu$ and $\Sigma$.

**Computational and data problems:**

**4**. In the baseball player data, run a linear regression of the Weight variable against Height, Age and Position. Note that the latter is a categorical variable, but as pointed out in the text, R will automatically create the proper dummy variables for you. Find a confidence interval for the difference in weights between pitchers and catchers, for fixed height and age.

**5**. In the baseball data, run a logistic regression, predicting Position from Height, Age and Weight.

**6**. Consider the Pima data, Section 7.8. Try predicting diabetic status from the other variables.

# Chapter 16

# Model Parsimony and Overfitting

We discussed model parsimony in Chapter 14. We now continue that discussion in the wake of what we now know about predictive modeling from the last chapter. Parsimony, for the sake of understanding and even aesthetics, is now eclipsed in importance by the related but separate problem of *overfitting*.

This is by far the most vexing issue in statistics and machine learning. The term refers to fitting too rich a model, given our sample size. Though it is most commonly discussed in regression problems, it is a general statistical issue.

## 16.1   What Is Overfitting?

### 16.1.1   Example: Histograms

We first saw this in Section 8.2.3.1. How many bins should one set in a histogram? If we have too few — in the extreme, just one — the basic character of the data will be lost. But with too many, we get a very jagged curve that seems to be "fitting the noise," again missing the fundamental trends in the data.

## 16.1.2    Example: Polynomial Regression

Suppose we have just one predictor, and $n$ data points. If we fit a polynomial model of degree $n - 1$, the resulting curve will pass through all $n$ points, a "perfect" fit. For instance:

```
> x <- rnorm(6)
> y <- rnorm(6) # unrelated to x!
> df <- data.frame(x,y)
> df$x2 <- x^2
> df$x3 <- x^3
> df$x4 <- x^4
> df$x5 <- x^5
> df
            x           y          x2           x3
1 -0.9808202   0.9898205  0.9620082  -0.94355703
2 -0.5115071   0.5725953  0.2616395  -0.13383047
3 -0.6824555  -0.1354214  0.4657456  -0.31785063
4  0.8113214   1.0621229  0.6582425   0.53404620
5 -0.7352366   0.1920320  0.5405728  -0.39744891
6  0.3600138   0.7356633  0.1296100   0.04666137
          x4          x5
1 0.92545975  -0.907709584
2 0.06845524  -0.035015341
3 0.21691892  -0.148037517
4 0.43328312   0.351531881
5 0.29221897  -0.214850074
6 0.01679874   0.006047779
> lmo <- lm(y ~ .,data=df)
> lmo$fitted.values
          1           2           3           4
  0.9898205   0.5725953  -0.1354214   1.0621229
          5           6
  0.1920320   0.7356633
> df$y
[1]  0.9898205   0.5725953  -0.1354214   1.0621229
[5]  0.1920320   0.7356633
```

The **fitted.values** component of an "lm" lists the values obtained when one uses the fitted model to predict the $Y$ values in the original dataset.

Yes, we "predicted" **y** perfectly, **even though there was no relation between the response and predictor variables).** Clearly that "perfect fit" is illusory, "noise fitting." Our ability to predict future cases would not be good. This is *overfitting*.

Let's take a closer look, say in an *recommender systems* context. A famous example involves predicting ratings moviegoers would give to various films. Say we are predicting a person's rating of a certain movie from previous ratings of other movies, plus the person's age and gender. Suppose men become more liberal raters as they age but women become more conservative. If we omit the interaction term, than we will underpredict older men and overpredict older women. This biases our ratings.

On the other hand, adding in the interaction terms may increase sampling variance, i.e., the standard errors of the estimated regression coeficients.

So we have the famous *bias/variance tradeoff*: As we use more and more terms in our regression model (predictors, polynomials, interaction terms), the bias decreases but the variance increases. This "tug of war" between these decreasing and increasing quantities typically yields a U-shaped curve: As we increase the number of terms from 1, mean absolute prediction error will at first decrease but eventually will increase. Once we get to the point at which it increases, we are *overfitting*.

This is particularly a problem when one has many dummy variables. For instance, there are more than 42,000 ZIP Codes in the US; to have a dummy for each would almost certainly be overfitting.

# 16.2  Can Anything Be Done about It?

So, where is the "happy medium," the model that is rich enough to capture most of the dynamics of the variables at hand, but simple enough to avoid variance issues? Unfortunately, **there is no good answer to this question.** We focus on the regression case here.

One quick rule of thumb is that one should have $p < \sqrt{n}$, where $p$ is the number of predictors, including polynomial and interaction terms, and $n$ is the number of cases in our sample. But this is certainly not a firm rule by any means.

## 16.2.1  Cross-Validation

From the polynomial-ftting example in Section 16.1, we see the following key point:

An assessment of predictive ability, based on predicting the same data on which our model is fit, tends to be overly optimistic and may be meaningless or close to it.

This motivates the most common approach to dealing with the bias/variance tradeoff, *cross validation*. In the simplest version, one randomly splits the data into a *training set* and a *test set*.[1] We fit the model to the training set and then, pretending we don't know the "Y" (i.e., response) values in the test set, predict those values from our fitted model and the "X" values (i.e., the predictors) in the test set. We then "unpretend," and check how well those predictions worked.

The test set is "fresh, new" data, since we called **lm()** or whatever only on the training set. Thus we are avoiding the "noise fitting" problem. We can try several candidate models — e.g., different sets of predictor variables or different numbers of nearest neighbors in a nonparametric setting — then choose the one that best predicts the test data.

Since the trainimg set/test set partitioning is random, we should perform the partitioning several times, thus assessing the performance of each of our candidate models several times to see if a clear pattern emerges.

(Note carefully that after fitting the model via cross-validation, we then use the full data for later prediction. Splitting the data for cross-validation was just a temporary device for model selection.)

Cross-validation is essentially the standard for model selection, and it works well if we only try a few models.

## 16.3   Predictor Subset Selection

So, we can use cross-validation to asses how well a particular set of predictors will do in new data. In principle, that means we could fit all possible subsets of our $p$ predictors, and choose the subset that does best in cross-validation. However, that may not work so well.

First, there are many, many subsets — $2^p$ of them! Even for moderately large $p$, evaluating all of them would be computationally infeasible.

Second, there could be a serious p-hacking problem (Section 10.16). In assessing hundreds or thousands of predictor subsets, it's quite possible that one of them accidentally looks promising.

---

[1]The latter is also called a *holdout set* or a *validation set*.

Unfortunately, there is no known effective solution to this dilemma, though a very large number of methods have been proposed. Many are in wide usage, including in R packages. We will not be able to pursue this further here.

## 16.4 Exercises

**Computational and data problems:**

**1.** Consider the **prgeng** dataset from Exercise 15, Chapter 11. You'll predict income from age, gender and occupation (dummy variables) and weeks worked. The relation between income and age is not linear, as income tends to level off after age 40 or so. So, try polynomial models for age. In other words, predict income from gender, occupation, weeks worked and age, age-squared, age-cubed on so on. See how high a polynomial can be fit before overfitting seems to occur. Assess using cross-validation, say with a test set of size 2500.

# Chapter 17

# Introduction to Discrete Time Markov Chains

In multivariate analysis, frequently one deals with *time series*, data that progresses through time. Weather, financial data, medical EEG/EKG tests and so on all fall into this category.

This is far too broad a topic for this book, so we will focus on one particular type of time series, *Markov chains*. In addition to its classic uses in areas such as queuing theory, genetics and physics, there are important data science applications, such as *Hidden Markov Models*, *Google PageRank* and *Markov Chain Monte Carlo*. We present the basic introductory ideas in this chapter.

The basic idea is that we have random variables $X_1, X_2, ...$, with the index representing time. Each one can take on any value in a given set, called the *state space*; $X_n$ is then the *state* of the system at time $n$. The state space is assumed either finite or *countably infinite*.[1] We sometimes also consider an initial state, $X_0$, which might be modeled as either fixed or random.

The key assumption is the *Markov property*, which in rough terms can be described as:

The probabilities of future states, given the present state and

---

[1]The latter is a mathematical term meaning, in essence, that it is possible to denote the space using integer subscripts. It can be shown that the set of all real numbers is not countably infinite, though perhaps surprisingly the set of all rational numbers *is* countably infinite.

the past states, depends only on the present state; the past is irrelevant.

In formal terms, the above prose is:

$$P(X_{t+1} = s_{t+1}|X_t = s_t, X_{t-1} = s_{t-1}, \ldots, X_0 = s_0)$$
$$= P(X_{t+1} = s_{t+1}|X_t = s_t) \qquad (17.1)$$

Note that in (17.1), the two sides of the equation are equal but their common value may depend on $t$. We assume that this is not the case; we assume nondependence on $t$, known as *stationarity*.[2] For instance, the probability of going from state 2 to state 5 at time 29 is assumed to be the same as the corresponding probability at time 333.

# 17.1   Matrix Formulation

We define $p_{ij}$ to be the probability of going from state $i$ to state $j$ in one time step; note that this is a *conditional* probability, i.e., $P(X_{n+1} = j \mid X_n = i)$. These quantities form a matrix $P$, whose row $i$, column $j$ element is $p_{ij}$, which is called the *transition matrix*.[3]

For example, consider a three-state Markov chain with transition matrix

$$P = \begin{pmatrix} \frac{1}{2} & 0 & \frac{1}{2} \\ \frac{1}{4} & \frac{1}{2} & \frac{1}{4} \\ 1 & 0 & 0 \end{pmatrix} \qquad (17.2)$$

This means for instance that if we are now at state 1, the probabilities of going to states 1, 2 and 3 are 1/2, 0 and 1/2, respectively. Note that each row's probabilities must sum to 1—after all, from any particular state, we must go *somewhere*.

Actually, the $m^{th}$ power, $P^m$, of the transition matrix gives the probabilities for $m$-step transitions. In other words, the $(i, j)$ element of $P^m$ is $P(X_{t+m} = j \mid X_t = i)$. This is clear for the case $m = 2$ as follows.

---

[2]Not to be confused with the notion of a stationary distribution, coming below.

[3]Unfortunately, we have some overloading of symbols here. Both in this book and in the field in general, we usually use the letter $P$ to denote this matrix, yet we continue to denote probabilities by $P()$. However, it will be clear from context which we mean. The same is true for our transition probabilities $p_{ij}$, which use a subscripted letter $p$, which is also the case for probability mass functions.

As usual, "break big events down into small events." How can it happen that $X_{t+2} = j$? Well, break things down according to where we might go first after leaving $i$. We might go from $i$ to 1, say, with probability $p_{i1}$, then from 1 to $j$, with probability $p_{1j}$. Similarly, we might go from $i$ to 2, then 2 to $j$, or from $i$ to 3 then 3 to $j$, etc. So,

$$P(X_{t+2} = j | X_t = i) = \sum_k p_{ik}\, p_{kj} \qquad (17.3)$$

In view of the rule for multiplying matrices, the expression on the right-hand side is simply the $(i, j)$ element of $P^2$!

The case of general $m$ then follows by mathematical induction.

# 17.2 Example: Die Game

Consider the following game. One repeatedly rolls a die, keeping a running total. Each time the total exceeds 10 (not equals 10), we receive one dollar, and continue playing, resuming where we left off, mod 10. Say for instance we have a cumulative total of 8, then roll a 5. We receive a dollar, and now our new total is 3.

It will simplify things if we assume that the player starts with one free point, i.e., $X_0 = 1$. We then never hit state 0, and thus can limit the state space to the numbers 1 through 10.

This process clearly satisfies the Markov property, with our state being our current total, 1, 2, ..., 10. If our current total is 6, for instance, then the probability that we next have a total of 9 is 1/6, *regardless of what happened in our previous rolls*. We have $p_{25}$, $p_{72}$ and so on all equal to 1/6, while for instance $p_{29} = 0$. Here's the code to find the transition matrix $P$:

```
# 10 states, so 10X10 matrix
# since most elements will be 0s,
# set them all to 0 first,
# then replace those that should be nonzero
p <- matrix(rep(0,100),nrow=10)
onesixth <- 1/6
for (i in 1:10) {  # look at each row
    # since we are rolling a die, there are
    # only 6 possible states we can go to
    # from i, so check these
```

```
for (j in 1:6) {
    k <- i + j  # new total, but did we win?
    if (k > 10) k <- k - 10
    p[i,k] <- onesixth
}
}
```

Note that since we knew that many entries in the matrix would be zero, it was easier just to make them all 0 first, and then fill in the nonzero ones. The initialization to 0 was done with the line

```
p <- matrix(rep(0,100),nrow=10)
```

See Section 17.8.1 for details.


## 17.3   Long-Run State Probabilities

In many applications of Markov modeling, our main interest is in the long-run behavior of the system. In particular, we are likely to visit some states more often than others, and wish to find the long-run probabilities of each state.

To that end, let $N_{it}$ denote the number of times we have visited state $i$ during times $1, ..., t$. For instance, in the die game, $N_{8,22}$ would be the number of rolls among the first 22 that resulted in our having a cumulative total of 8.

In typical applications we have that the proportion

$$\pi_i = \lim_{t \to \infty} \frac{N_{it}}{t} \tag{17.4}$$

exists for each state $i$, regardless of where we start. Under a couple more conditions,[4] we have the stronger result,

$$\lim_{t \to \infty} P(X_t = i) = \pi_i \tag{17.5}$$

---

[4]Basically, we need the chain to not be *periodic*. Consider a random walk, for instance: We start at position 0 on the number line, at time 0. The states are the integers. (So, this chain has an infinite state space.) At each time, we flip a coin to determine whether to move right (heads) or left (tails) 1 unit. A little thought shows that if we start at 0, the only times we can return to 0 are even-numbered times, i.e., $P(X_n = 0 \,|\, X_0 = 0)$ for all odd numbers $n$. This is a periodic chain. By the way, (17.4) turns out to be 0 for this chain.

These quantities $\pi_i$ are typically the focus of analysis of Markov chains.
We will use the symbol $\pi$ to name the column vector of all the $\pi_i$:

$$\pi = (\pi_1, \pi_2, ...)' \tag{17.6}$$

where $'$ as usual means matrix transpose.

## 17.3.1   Stationary Distribution

The $\pi_i$ are called *stationary probabilities*, because if the initial state $X_0$
is a random variable with that distribution, then all $X_i$ will have that
distribution. Here's why:

Assuming (17.5), we have

$$\pi_i \;=\; \lim_{n \to \infty} P(X_n = i) \tag{17.7}$$

$$=\; \lim_{n \to \infty} \sum_k P(X_{n-1} = k)\, p_{ki} \tag{17.8}$$

$$=\; \sum_k \pi_k\, p_{ki} \tag{17.9}$$

(The summation in the second equation reflects our usual question in probability problems, "How can it happen?" In this case, we break the event
$X_n = i$ down according to where we might have been at time $n - 1$.)

In summary, for each $i$ we have

$$\pi_i = \sum_k \pi_k\, p_{ki} \tag{17.10}$$

Usually we take $X_0$ to be a constant. But let's allow it to be a random
variable, with distribution $\pi$, i.e., $P(X_0 = i) = \pi_i$. Then

$$P(X_1 = i) \;=\; \sum_k P(X_0 = k)\, p_{ki} \tag{17.11}$$

$$=\; \sum_k \pi_k\, p_{ki} \tag{17.12}$$

$$=\; \pi_i \tag{17.13}$$

this last using (17.10). So, if $X_0$ has distribution $\pi$, then the same will be true for $X_1$, and continuing in this manner we see that $X_2, X_3, \ldots$ will all have that distribution, thus demonstrating the claimed stationary property for $\pi$.

Of course, (17.10) holds for all states $i$. The reader should verify that in matrix terms, (17.10) says

$$\pi' = \pi'P \qquad\qquad (17.14)$$

For instance, for a 3-state chain, the first column of $P$ will be $(p_{11}, p_{21}, p_{31})'$. Multiplying this on the left by $\pi' = (\pi_1, \pi_2, \pi_3)$, we get

$$\pi_1 p_{11} + \pi_2 p_{21} + \pi_3 p_{31} \qquad\qquad (17.15)$$

which by (17.10) is $\pi_1$, just as claimed by (17.14).

This equation turns out to be key to actually calculating the vector $\pi$, as we will now see.

## 17.3.2   Calculation of $\pi$

Equation (17.14) then shows us how to find the $\pi_i$, at least in the case of finite state spaces, the subject of this section here, as follows.

First, rewrite (17.14)

$$(I - P')\pi = 0 \qquad\qquad (17.16)$$

Here $I$ is the $n \times n$ identity matrix (for a chain with $n$ states).

This equation has infinitely many solutions; if $\pi$ is a solution, then so for example is $8\pi$. Moreover, the equation shows that the matrix there, $I - P'$, cannot have an inverse; if it did, we could multiply both sides by the inverse, and find that the unique solution is $\pi = 0$, which can't be right. Linear algebra theory in turn implies that the rows of $I - P'$ are not linearly independent; in plain English, at least one of those equations is redundant.

But we need $n$ independent equations, and fortunately an $n^{th}$ one is available:

$$\sum_i \pi_i = 1 \qquad (17.17)$$

Note that (17.17) can be written as

$$O'\pi = 1 \qquad (17.18)$$

where $O$ ("one") is a vector of $n$ 1s. Excellent, let's use it!

So, again, thinking of (17.16) as a system of linear equations, let's replace the last equation by (17.18). Switching back to the matrix view, that means that we replace the last row in the matrix $I - P'$ in (17.16) by $O'$, and correspondingly replace the last element of the right-side vector by 1. Now we have a nonzero vector on the right side, and a full-rank (i.e., invertible) matrix on the left side. This is the basis for the following code, which we will use for finding $\pi$.

```
findpi1 <- function(p) {
   n <- nrow(p)
   # find I-P'
   imp <- diag(n) - t(p)   # diag(n) = I, t() = '
   # replace the last row of I-P' as discussed
   imp[n,] <- rep(1,n)
   # replace the corresponding element of the
   # right side by (the scalar) 1
   rhs <- c(rep(0,n-1),1)
   # now use R's built-in solve()
   solve(imp,rhs)
}
```

### 17.3.3   Simulation Calculation of $\pi$

In some applications, the state space is huge. Indeed, in the case of Google PageRank, there is a state for each Web page, thus a state space running to the hundreds of millions! So the matrix solution above is infeasible. In this case, a simulation approach can be helpful, as follows.

Recall our comments in Section 17.3:

...let $N_{it}$ denote the number of times we have visited state $i$ during times $1, ..., t$. In typical applications we have that

$$\pi_i = \lim_{t \to \infty} \frac{N_{it}}{t}$$

exists for each state $i$, **regardless of where we start**.

So, we can choose an initial state, and then simulate the action of the chain for a number of time steps, and then report the proportion of time we spent at each state. That will give us the approximate value of $\pi$.

Below is code implementing this idea. The arguments are **p**, our transition matrix; **nsteps**, the length of time we wish to run the simulation; and **x0**, our chosen initial state.

```
simpi <- function(p,nsteps,x0)
{
    nstates <- ncol(p)
    visits <- rep(0,nstates)
    x <- x0
    for (step in 1:nsteps) {
        x <- sample(1:nstates,1,prob=p[x,])
        visits[x] <- visits[x] + 1
    }
    visits / nsteps
}
```

The vector **visits** keeps track of how often we've been to each state. When we are at state **x**, we randomly choose our next state according to the chain's transition probabilities from that state:

```
x <- sample(1:nstates,1,prob=p[x,])
```

Note that at the outset of this section, it was said that the state space can be huge. In fact, it can in principle be infinite, say for a queuing system with unlimited buffer space. In such cases, the above code would still work, with some modification (Exercise 7).

# 17.4   Example: 3-Heads-in-a-Row Game

How about the following game? We keep tossing a coin until we get three consecutive heads. What is the expected value of the number of tosses we need?[5]

We can model this as a Markov chain with states 0, 1, 2 and 3, where state $i$ means that we have accumulated $i$ consecutive heads so far. Let's model the game as being played repeatedly, as in the die game above. Note that now that we are taking that approach, it will suffice to have just three states, 0, 1 and 2; there is no state 3, because as soon as we win, we immediately start a new game, in state 0.

Clearly we have transition probabilities such as $p_{01}$, $p_{12}$, $p_{10}$ and so on all equal to 1/2. Note from state 2 we can only go to state 0, so $p_{20} = 1$.

Below is the code to set the matrix $P$ and solve for $\pi$. Of course, since R subscripts start at 1 instead of 0, we must recode our states as 1, 2 and 3.

```
p <- matrix(rep(0,9),nrow=3)
p[1,1] <- 0.5
p[1,2] <- 0.5
p[2,3] <- 0.5
p[2,1] <- 0.5
p[3,1] <- 1
findpi1(p)
```

It turns out that

$$\pi = (0.5714286, 0.2857143, 0.1428571) \qquad (17.19)$$

So, in the long run, about 57.1% of our tosses will be done while in state 0, 28.6% while in state 1, and 14.3% in state 2.

Now, look at that latter figure. Of the tosses we do while in state 2, half will be heads, so half will be wins. In other words, about 0.071 of our tosses will be wins. And THAT figure answers our original question (expected number of tosses until win), through the following reasoning:

Think of, say, 10000 tosses. There will be about 710 wins sprinkled among those 10000 tosses. Thus the average number of tosses between wins will be about $10000/710 = 14.1$. In other words, the expected time until we get three consecutive heads is about 14.1 tosses.

---

[5]By the way, this was actually an interview question given to an applicant for a job as a Wall Street "quant," i.e., quantitative modeler.

# 17.5   Example: Bus Ridership Problem

Consider the bus ridership problem in Section 1.1. Make the same assumptions now, but add a new one: There is a maximum capacity of 20 passengers on the bus. (Assume all alighting passengers leave before any passengers are allowed to board.)

It may be helpful to review the notation and assumptions:

- $L_i$: number of passengers on the bus as it *leaves* stop $i$

- $B_i$: number of passengers who board the bus at stop $i$

- At each stop, each passssenger alights from the bus with probability 0.2.

- At each stop, either 0, 1 or 2 new passengers get on the bus, with probabilities 0.5, 0.4 and 0.1, respectively.

We will also define:

- $G_i$: number of passengers who get off the bus at stop $i$

The random variables $L_i$, $i = 1, 2, 3, ...$ form a Markov chain. Let's look at some of the transition probabilities:

$$p_{00} = 0.5$$

$$p_{01} = 0.4$$

$$p_{11} = (1 - 0.2) \cdot 0.5 + 0.2 \cdot 0.4$$

$$p_{20} = (0.2)^2(0.5) = 0.02$$

$$p_{20,20} = (0.8)^{20}(0.5 + 0.4 + 0.1)+$$
$$\binom{20}{1}(0.2)^1(0.8)^{20-1}(0.4 + 0.1) + \binom{20}{2}(0.2)^2(0.8)^{18}(0.1)$$

(Note that for clarity, there is a comma in $p_{20,20}$, as $p_{2020}$ would be confusing and in some other examples even ambiguous. A comma is not necessary in $p_{11}$, since there must be two subscripts; the 11 here can't be eleven.)

After finding the $\pi$ vector as above, we can find quantities such as the long-run average number of passengers on the bus,

$$\sum_{i=0}^{20} \pi_i i \tag{17.20}$$

We can also compute the long-run average number of would-be passengers who fail to board the bus. Denote by $A_i$ the number of passengers on the bus as it *arrives* at stop $i$. The key point is that since $A_i = L_{i-1}$, then (17.4) and (17.5) will give the same result, no matter whether we look at the $L_j$ chain or the $A_j$ chain.

Now, armed with that knowledge, let $D_j$ denote the number of disappointed people at stop $j$, i.e., the number who fail to board the bus. Then

$$ED_j = 1 \cdot P(D_j = 1) + 2 \cdot P(D_j = 2). \tag{17.21}$$

That latter probability, for instance, is

$$
\begin{aligned}
P(D_j = 2) &= P(A_j = 20 \text{ and } B_j = 2 \text{ and } G_j = 0) \\
&= P(A_j = 20) \, P(B_j = 2) \, P(G_j = 0 \mid A_j = 20) \\
&= P(A_j = 20) \cdot 0.1 \cdot 0.8^{20}
\end{aligned}
$$

Using the same reasoning, one can find $P(D_j = 1)$. (A number of cases to consider, left as an exercise for the reader.)

Taking limits in (17.21) as $j \to \infty$, we have the long-run average number of disappointed customers on the left, and on the right, the term $P(A_j = 20)$ goes to $\pi_{20}$, which we have and thus can obtain the value regarding disappointed customers.

## 17.6 Hidden Markov Models

Though Markov models are a mainstay in classical applied mathematics, they are the basis for a very interesting more recent application tool known to data science known as *Hidden Markov Models* (HMMs). The fields of interest are those with sequential data, such as text classification; the latter is sequential since the words in a text come in sequence. Speech recognition is similarly sequential, as is genetic code analysis and so on.

The Markov chains in such models are "hidden" in that they are not observed. In text processing, we observe the words but not the underlying grammatical parts-of-speech, which we might model as Markovian.

The goal of HMMs is to guess the hidden states for the Markov chain from the information we have on observable quantities. We find the most likely sequence in the former and use that as our guess.

## 17.6.1   Example: Bus Ridership

A parts-of-speech model would be too complex to present here, but our familiar bus ridership model will illustrate HMM concepts well.

Recall that $L_1, L_2, ...$ form a Markov chain. But suppose that we do not actually have data on the number of people on the bus, and all we know is the number of people $G_1$, $G_2, ...$ exiting the bus, say observing them come through a gate some distance away from the bus stop. We know the bus is empty as it arrives at the first stop, and thus that $G_1 = 0$.

We'll keep our example small, but it should be able to capture the spirit of HMM. Say we observe the bus for just one stop, the second. So we wish to guess $L_1$ and $L_2$ based on $G_2$. For instance, say we observe $G_2 = 0$. Here are a few of the possibilities and their probabilities:

- $B_1 = 0$, $B_2 = 0$, $G_2 = 0$: $L_1 = 0$, $L_2 = 0$, probability $0,5^2 \cdot 1 = 0.25$

- $B_1 = 1$, $B_2 = 0$, $G_2 = 0$: $L_1 = 1$, $L_2 = 1$, probability $0.4 \cdot 0,5 \cdot 0.8 = 0.16$

- $B_1 = 1$, $B_2 = 1$, $G_2 = 0$: $L_1 = 1$, $L_2 = 2$, probability $0.4^2 \cdot 0.8 = 0.128$

- etc.

After tabulating all the possibilities, we would take as our guess for $L_1$ and $L_2$ the one with the largest probability. In other words, we are doing Maximum Likelihood Estimation, having observed $G_2 = 0$. Among the first three above, that would be $L_1 = 0$, $L_2 = 0$, though of course there are many other cases not checked yet.

## 17.6.2 Computation

Even in our very small example above, there were many cases to enumerate. Clearly, it is difficult to keep track of all the possible cases in large applications, and there is an issue of doing so efficiently. Fortunately, efficient algorithms have been developed for this and implemented in software, including in R. See for instance [27].

# 17.7 Google PageRank

As the reader probably knows, Google's initial success was spurred by the popularity of its search engine, PageRank. The name is a pun, alluding both to the fact that the algorithm involves Web *pages*, and to the surname of its inventor, Google cofounder Larry *Page*.[6]

The algorithm models the entire Web as a huge Markov chain. Each Web page is a state in the chain, and the transition probabilities $p_{ij}$ are the probabilities that a Web surfer currently at site $i$ will next visit site $j$.

One might measure the popularity of site $k$ by computing the stationary probability $\pi_k$. PageRank essentially does this, but adds weights to the model that further accentuate the popularity of top sites.

# 17.8 Computational Complements

## 17.8.1 Initializing a Matrix to All 0s

In the code in Section 17.2, we found it convenient to first set all elements of the matrix to 0. This was done by the line

```
p <- matrix(rep(0,100),nrow=10)
```

R's **rep()** ("repeat") function does what the name implies. Here it repeats the value zero 100 times. We have a $10 \times 10$ matrix, so we do need 100 zeros.

Note that R uses *column-major* order for matrix storage in memory: First all of column 1 is stored, then all of column 2, and so on. So for instance:

---

[6]Page holds the patent, though mentions benefiting from conversations with several others, including fellow Google cofounder Sergei Brin.

```
> matrix(c(5,1,8,9,15,3),ncol=2)
      [,1] [,2]
[1,]    5    9
[2,]    1   15
[3,]    8    3
```

## 17.9  Exercises

**Mathematical problems:**

**1.** In the bus ridership example, Section 17.5, find $p_{31}$ and $p_{19,18}$.

**2.** Consider the 3-heads-in-a-row game, Section 17.4. Modify the analysis for a 2-heads-in-a-row game.

**3.** Consider a game in which the player repeatedly rolls a die. Winning is defined to be have rolled at least one 1 and at least one 2. Find the expected time to win.

**4.** In the die game, Section 17.2, use Markov analysis to find the mean time between wins.

**5.** Consider the die game, Section 17.2. Suppose on each turn we roll two dice instead of one. We still have 10 states , 1 through 10, but the transition matrix $P$ changes. If for instance we are now in state 6 and roll (3,2), we win a dollar and our next state is 1. Find the first row of $P$.

**6.** Consider the 3-heads-in-a-row game, Section 17.4, a Markov chain with states 0, 1 and 2. Let $W_i$ denote the time it takes to next reach state 2, starting in state $i$, $i = 0, 1, 2$. (Note the word "next"; $W_2$ is not 0.) Let $d_i = EW_i, i = 0, 1, 2$. Find the vector $(d_0, d_1, d_2)$.

**Computational and data problems:**

**7.** Implement the suggestion at the end of Section 17.3.3. Replace the matrix **p** by a function **p()**, and have **visits** keep track only of states visited so far, say in an R **list**.

**8.** Write an R function with call form **calcENit(P,s,i,t)** to calculate $EN_{it}$ in (17.4), starting from state **s**. Hint: Use indicator random variables.

**9.** In our study of Markov chains here, the transition matrix $P$ was always either given to us, or easily formulated from the physical structure of the problem. In some applications, though, we may need to estimate $P$ from data.

Consider for example the Nile dataset that is built in to R. We might try to fit a Markov model. To make the state space discrete, write a function **makeState(x,nc)**, which splits the input time series **x** into **nc** equal intervals. It is suggested that you use R's **cut()** function for this.

Then write a function **findP(xmc)** to estimate $P$; here **xmc** is the output of **makeState()**.

One could go further. Instead of defining a state as a single point in the time series, we could define it to be a pair of consecutive points, thus having a more general type of dependency.

# Part IV

# Appendices

# Appendix A

# R Quick Start

Here we present a quick introduction to the R data/statistical programming language. Armed with this material, the reader will be well equipped to read, understand and use advanced material in one of the many Web tutorials. Or in book form, there is my book, [28]. For a more advanced level, see [43], and for internal details, there is [6].

It is assumed here that the reader has some prior experience with Python or C/C++, meaning e.g., that he/she is familiar with loops and "if/else."[1]

Python users will find it especially easy to become adept at R, as they both have a very useful interactive mode. Readers should follow my motto, "When in doubt, try it out!" In interactive mode, one can try quick little experiments to learn/verify how R constructs work.

## A.1  Starting R

To invoke R, just type "R" into a terminal window, or click an icon if your desktop has one.

If you prefer to run from an IDE, you may wish to consider ESS for Emacs, StatET for Eclipse or RStudio, all open source. ESS is the favorite among the "hard core coder" types, while the colorful, easy-to-use, RStudio is a

---

[1]For readers who happen to be computer science specialists: R is object-oriented (in the sense of *encapsulation, polymorphism* and everything being an object) and is a functional language (i.e., almost no side effects, every action is a function call, etc. For example, the expression **2+5** implemented as a function call, "+"(2,5).

big general crowd pleaser. If you are already an Eclipse user, StatET will be just what you need.[2]

R is normally run in interactive mode, with > as the prompt. For batch work, use **Rscript**, which is in the R package.

# A.2   Correspondences

Here is how Python, C/C++ and R compare in terms of:

- assignment operator
- array terminology
- subscripts/indexes
- 2-D array notation
- 2-D array storage
- mixed container type
- mechanism for external code packaging
- run mode
- comment symbol

| Python | C/C++ | R |
|---|---|---|
| = | = | <- (or =) |
| list | array | vector, matrix, array |
| start at 0 | start at 0 | start at 1 |
| m[2][3] | m[2][3] | m[2,3] |
| NA | row-major order | column-major order |
| dictionary | struct | list |
| import | include, link | library() |
| interactive, batch | batch | interactive, batch |
| # | // | # |

---

[2]I personally use **vim**, as I want to have the same text editor no matter what kind of work I am doing. But I have my own macros to help with R work.

# A.3    First Sample Programming Session

Below is a commented R session, to introduce the concepts. I had a text editor open in another window, constantly changing my code, then loading it via R's **source()** command. The original contents of the file **odd.R** were:

```
oddcount <- function(x)   {
    k <- 0   # assign 0 to k
    for (n in x)   {   # loop through all of x
        if (n %% 2 == 1)   # n odd?
            k <- k+1
    }
    return(k)
}
```

The function counts the number of odd numbers in a vector **x**. We test for a number **n** being odd by using the "mod" operator, which calculates remainders upon division. For instance, 29 mod 7 is 1, since 29 divided by 7 is 4 with a remainder of 1. To check whether a number is odd, which determine whether its value mod 2 is 1.

By the way, we could have written that last statement as simply

```
    k
```

because the last computed value of an R function is returned automatically. This is actually preferred style in the R community.

The R session is shown below. You may wish to type it yourself as you go along, trying little experiments of your own along the way.

```
> source("odd.R")   # load code from the given file
> ls()   # what objects do we have?
[1] "oddcount"

# what kind of object is oddcount (well,
# we already know)?

> class(oddcount)
[1] "function"

# while in interactive mode, and not inside
# a function, can print any object by typing
# its name; otherwise use print(), e.g., print(x+y)
```

```
> oddcount   # function is object, so can print it
function(x)  {
   k <- 0  # assign 0 to k
   for (n in x)  {
      if (n %% 2 == 1) k <- k+1
   }
   return(k)
}

# let's test oddcount(), but look at some
# properties of vectors first

> y <- c(5,12,13,8,88)   # the concatenate function
> y
[1]   5 12 13   8 88
> y[2]   # R subscripts begin at 1, not 0
[1] 12
> y[2:4]   # extract elements 2, 3 and 4 of y
[1] 12 13   8
> y[c(1,3:5)]   # elements 1, 3, 4 and 5
[1]   5 13   8 88
> oddcount(y)   # should report 2 odd numbers
[1] 2

# change code (in the other window) to vectorize
# the count operation, for much faster execution

> source("odd.R")
> oddcount
function(x)  {
   x1 <- (x %% 2 == 1)
   # x1 now a vector of TRUEs and FALSEs
   x2 <- x[x1]
   # x2 now has the elements of x that
   # were TRUE in x1
   return(length(x2))
}

# try it on subset of y, elements 2 through 3

> oddcount(y[2:3])
```

```
[1]  1
# try it on subset of y, elements 2, 4 and 5

> oddcount(y[c(2,4,5)])
[1]  0

> # further compactify the code
> source("odd.R")
> oddcount
function(x)   {
   length(x[x %% 2 == 1])
      # last value computed is auto returned
}
> oddcount(y)   # test it
[1]  2
```

```
# and even more compactification, making
# use of the fact that TRUE and
# FALSE are treated as 1 and 0
```

```
> oddcount <- function(x) sum(x %% 2 == 1)
```

```
# make sure you understand the steps that
# that involves:  x is a vector, and thus
# x %% 2 is a new vector, the result of
# applying the mod 2 operation to every
# element of x; then x %% 2 == 1 applies
# the == 1 operation to each element of
# that result, yielding a new vector of
# TRUE and FALSE values; sum() then adds
# them (as 1s and 0s)
```

```
# we can also determine which elements are odd
```

```
> which(y %% 2 == 1)
[1]  1 3
```

Note that, as I like to say, "the function of the R function **function()** is to produce functions!" Thus assignment is used. For example, here is what **odd.R** looked like at the end of the above session:

```
oddcount <- function(x)   {
   x1 <- x[x %% 2 == 1]
```

```
    return(list(odds=x1, numodds=length(x1)))
}
```

We created some code, and then used **function()** to create a function object, which we assigned to **oddcount**.

## A.4   Vectorization

Note that we eventually *vectorized* our function **oddcount()**. This means taking advantage of the vector-based, functional language nature of R, exploiting R's built-in functions instead of loops. This changes the venue from interpreted R to C level, with a potentially large increase in speed. For example:

```
> x <- runif(1000000)   # 10^6 random nums from (0,1)
> system.time(sum(x))
   user   system elapsed
  0.008    0.000    0.006
> system.time({s <- 0;
      for (i in 1:1000000) s <- s + x[i]})
   user   system elapsed
  2.776    0.004    2.859
```

## A.5   Second Sample Programming Session

A matrix is a special case of a vector, with added class attributes, the numbers of rows and columns.

```
# rbind() function combines rows of matrices;
# there's a cbind() too

> m1 <- rbind(1:2,c(5,8))
> m1
     [,1] [,2]
[1,]    1    2
[2,]    5    8
> rbind(m1,c(6,-1))
     [,1] [,2]
[1,]    1    2
[2,]    5    8
[3,]    6   -1
```

```
# form matrix from 1,2,3,4,5,6, in 2 rows

> m2 <- matrix(1:6,nrow=2)
> m2
     [,1] [,2] [,3]
[1,]    1    3    5
[2,]    2    4    6
> ncol(m2)
[1] 3
> nrow(m2)
[1] 2
> m2[2,3]   # extract element in row 2, col 3
[1] 6
# get submatrix of m2, cols 2 and 3, any row

> m3 <- m2[,2:3]
> m3
     [,1] [,2]
[1,]    3    5
[2,]    4    6


# or write to that submatrix

> m2[,2:3] <- cbind(c(5,12),c(8,0))
> m2
     [,1] [,2] [,3]
[1,]    1    5    8
[2,]    2   12    0


> m1 * m3   # elementwise multiplication
     [,1] [,2]
[1,]    3   10
[2,]   20   48
> 2.5 * m3  # scalar multiplication (but see below)
     [,1] [,2]
[1,]  7.5 12.5
[2,] 10.0 15.0
> m1 %*% m3  # linear algebra matrix multiplication
     [,1] [,2]
[1,]   11   17
[2,]   47   73
```

```
# matrices are special cases of vectors,
# so can treat them as vectors

> sum(m1)
[1] 16
> ifelse(m2 %%3 == 1,0,m2) # (see below)
     [,1] [,2] [,3]
[1,]    0    3    5
[2,]    2    0    6
```

## A.6    Recycling

The "scalar multiplication" above is not quite what you may think, even though the result may be. Here's why:

In R, scalars don't really exist; they are just one-element vectors. However, R usually uses *recycling*, i.e., replication, to make vector sizes match. In the example above in which we evaluated the express 2.5 * m3, the number 2.5 was recycled to the matrix

$$\begin{pmatrix} 2.5 & 2.5 \\ 2.5 & 2.5 \end{pmatrix} \tag{A.1}$$

in order to conform with **m3** for (elementwise) multiplication.

## A.7    More on Vectorization

The **ifelse()** function is another example of vectorization. Its call has the form

```
ifelse(boolean vectorexpression1, vectorexpression2,
    vectorexpression3)
```

All three vector expressions must be the same length, though R will lengthen some via recycling. The action will be to return a vector of the same length (and if matrices are involved, then the result also has the same shape). Each element of the result will be set to its corresponding element in **vectorexpression2** or **vectorexpression3**, depending on whether the corresponding element in **vectorexpression1** is TRUE or FALSE.

In our example above,

```
> ifelse(m2 %%3 == 1,0,m2) # (see below)
```

the expression m2 %%3 == 1 evaluated to the boolean matrix

$$
\begin{pmatrix}
T & F & F \\
F & T & F
\end{pmatrix}
\tag{A.2}
$$

(TRUE and FALSE may be abbreviated to T and F.)

The 0 was recycled to the matrix

$$
\begin{pmatrix}
0 & 0 & 0 \\
0 & 0 & 0
\end{pmatrix}
\tag{A.3}
$$

while **vectorexpression3**, **m2**, evaluated to itself.

# A.8 Default Argument Values

Consider the **sort()** function, which is built-in to R, though the following points hold for any function, including ones you write yourself.

The online help for this function, invoked by

```
> ?sort
```

shows that the call form (the simplest version) is

```
sort(x, decreasing = FALSE, ...)
```

Here is an example:

```
> x <- c(12,5,13)
> sort(x)
[1]  5 12 13
> sort(x,decreasing=FALSE)
[1] 13 12  5
```

So, the default is to sort in ascending order, i.e., the argument **decreasing** has TRUE as its default value. If we want the default, we need not specify this argument. If we want a descending-order sort, we must say so.

# A.9   The R List Type

The R **list** type is, after vectors, the most important R construct. A list is like a vector, except that the components are generally of mixed types.

## A.9.1   The Basics

Here is example usage:

```
> g <- list(x = 4:6, s = "abc")
> g
$x
[1] 4 5 6

$s
[1] "abc"

> g$x   # can reference by component name
[1] 4 5 6
> g$s
[1] "abc"
> g[[1]]   # can ref. by index; note double brackets
[1] 4 5 6
> g[[2]]
[1] "abc"
> for (i in 1:length(g)) print(g[[i]])
[1] 4 5 6
[1] "abc"

# now have ftn oddcount() return odd count
# AND the odd numbers themselves, using the
# R list type

> source("odd.R")
> oddcount
function(x)  {
   x1 <- x[x %% 2 == 1]
   list(odds=x1, numodds=length(x1))
}
> # R's list type can contain any type;
> #components delineated by $
> oddcount(y)
```

```
$odds
[1]   5 13

$numodds
[1]  2

> ocy <- oddcount(y)
> ocy
$odds
[1]   5 13

$numodds
[1]  2

> ocy$odds
[1]   5 13
> ocy[[1]]    # can get list elts. using [[ ]] or $
[1]   5 13
> ocy[[2]]
[1]  2
```

## A.9.2   S3 Classes

R is an object-oriented (and functional) language. It features two types of classes (actually more), S3 and S4. I'll introduce S3 here.

An S3 object is simply a list, with a class name added as an *attribute*:

```
> j <- list(name="Joe", salary=55000, union=T)
> class(j) <- "employee"
> m <- list(name="Joe", salary=55000, union=F)
> class(m) <- "employee"
```

So now we have two objects of a class we've chosen to name **"employee"**. Note the quotation marks.

We can write class *generic functions* (Section 8.9.1):

```
> print.employee <- function(wrkr) {
+     cat(wrkr$name,"\n")
+     cat("salary",wrkr$salary,"\n")
+     cat("union member",wrkr$union,"\n")
+ }
```

```
> print(j)
Joe
salary 55000
union member  TRUE
> j
Joe
salary 55000
union member  TRUE
```

What just happened?  Well, **print()** in R is a *generic* function, meaning that it is just a placeholder for a function specific to a given class.  When we printed **j** above, the R interpreter searched for a function **print.employee()**, which we had indeed created, and that is what was executed.  Lacking this, R would have used the print function for R lists, as before:

```
> rm(print.employee)
> # remove function, see what happens with print
> j
$name
[1] "Joe"

$salary
[1] 55000

$union
[1] TRUE

attr(,"class")
[1] "employee"
```

## A.10    Data Frames

Another workhorse in R is the *data frame*.  A data frame works in many ways like a matrix, but differs from a matrix in that it can mix data of different modes.  One column may consist of integers, while another can consist of character strings and so on.  Within a column, though, all elements must be of the same mode, and all columns must have the same length.

We might have a 4-column data frame on people, for instance, with columns for height, weight, age and name—3 numeric columns and 1 character string column.

Technically, a data frame is an R list, with one list element per column; each column is a vector. Thus columns can be referred to by name, using the **$** symbol as with all lists, or by column number, as with matrices. The matrix **a[i,j]** notation for the element of **a** in row **i**, column **j**, applies to data frames. So do the **rbind()** and **cbind()** functions, and various other matrix operations, such as filtering.

Here is an example using the dataset **airquality**, built in to R for illustration purposes. You can learn about the data through R's online help, i.e.,

```
> ?airquality
```

Let's try a few operations:

```
> names(airquality)
[1] "Ozone"   "Solar.R" "Wind"    "Temp"    "Month"
"Day"
> head(airquality)  # look at the first few rows
  Ozone Solar.R Wind Temp Month Day
1    41     190  7.4   67     5   1
2    36     118  8.0   72     5   2
3    12     149 12.6   74     5   3
4    18     313 11.5   62     5   4
5    NA      NA 14.3   56     5   5
6    28      NA 14.9   66     5   6
> airquality[5,3]  # wind on the 5th day
[1] 14.3
> airquality$Wind[3]  # same
[1] 12.6
> nrow(airquality)  # number of days observed
[1] 153
> ncol(airquality)  # number of variables
[1] 6
> airquality$Celsius <-
    (5/9) * (airquality[,4] - 32)  # new column
> names(airquality)
[1] "Ozone"   "Solar.R" "Wind"    "Temp"    "Month"
"Day"      "Celsius"
> ncol(airquality)
[1] 7
> airquality[1:3,]
  Ozone Solar.R Wind Temp Month Day  Celsius
1    41     190  7.4   67     5   1 19.44444
```

```
2     36      118  8.0    72       5    2 22.22222
3     12      149 12.6    74       5    3 23.33333
# filter op
> aqjune <- airquality[airquality$Month == 6,]
> nrow(aqjune)
[1] 30
> mean(aqjune$Temp)
[1] 79.1
# write data frame to file
> write.table(aqjune,"AQJune")
> aqj <- read.table("AQJune",header=T)   # read it in
```

## A.11   Online Help

R's **help()** function, which can be invoked also with a question mark, gives short descriptions of the R functions. For example, typing

```
> ?rep
```

will give you a description of R's **rep()** function.

An especially nice feature of R is its **example()** function, which gives nice examples of whatever function you wish to query. For instance, typing

```
> example(wireframe())
```

will show examples — R code and resulting pictures — of **wireframe()**, one of R's 3-dimensional graphics functions.

## A.12   Debugging in R

The internal debugging tool in R, **debug()**, is usable but rather primitive. Here are some alternatives:

- The RStudio IDE has a built-in debugging tool.

- For Emacs users, there is **ess-tracebug**.

- The StatET IDE for R on Eclipse has a nice debugging tool. Works on all major platforms, but can be tricky to install.

- My own debugging tool, **dbgR**, is extensive and easy to install, but for the time being is limited to Linux, Mac and other Unix-family systems. See *http://github.com/matloff/dbgR*.

# Appendix B

# Matrix Algebra

This appendix is intended as a review of basic matrix algebra, or a quick treatment for those lacking this background.

## B.1   Terminology and Notation

A *matrix* is a rectangular array of numbers. A *vector* is a matrix with only one row (a *row vector*) or only one column (a *column vector*).

The expression, "the $(i, j)$ element of a matrix," will mean its element in row $i$, column $j$.

If $A$ is a *square* matrix, i.e., one with equal numbers $n$ of rows and columns, then its *diagonal* elements are $a_{ii}$, $i = 1, ..., n$.

### B.1.1   Matrix Addition and Multiplication

- For two matrices have the same numbers of rows and same numbers of columns, addition is defined elementwise, e.g.,

$$
\begin{pmatrix} 1 & 5 \\ 0 & 3 \\ 4 & 8 \end{pmatrix} + \begin{pmatrix} 6 & 2 \\ 0 & 1 \\ 4 & 0 \end{pmatrix} = \begin{pmatrix} 7 & 7 \\ 0 & 4 \\ 8 & 8 \end{pmatrix} \tag{B.1}
$$

- Multiplication of a matrix by a *scalar*, i.e., a number, is also defined

383

elementwise, e.g.,

$$0.4 \begin{pmatrix} 7 & 7 \\ 0 & 4 \\ 8 & 8 \end{pmatrix} = \begin{pmatrix} 2.8 & 2.8 \\ 0 & 1.6 \\ 3.2 & 3.2 \end{pmatrix} \tag{B.2}$$

- The *inner product* or *dot product* of equal-length vectors $X$ and $Y$ is defined to be

$$\sum_{k=1}^{n} x_k y_k \tag{B.3}$$

- The product of matrices $A$ and $B$ is defined if the number of rows of $B$ equals the number of columns of $A$ ($A$ and $B$ are said to be *conformable*).  In that case, the $(i, j)$ element of the product $C$ is defined to be

$$c_{ij} = \sum_{k=1}^{n} a_{ik} b_{kj} \tag{B.4}$$

For instance,

$$\begin{pmatrix} 7 & 6 \\ 0 & 4 \\ 8 & 8 \end{pmatrix} \begin{pmatrix} 1 & 6 \\ 2 & 4 \end{pmatrix} = \begin{pmatrix} 19 & 66 \\ 8 & 16 \\ 24 & 80 \end{pmatrix} \tag{B.5}$$

It is helpful to visualize $c_{ij}$ as the inner product of row $i$ of $A$ and column $j$ of $B$, e.g., as shown in bold face here:

$$\begin{pmatrix} \mathbf{7} & \mathbf{6} \\ 0 & 4 \\ 8 & 8 \end{pmatrix} \begin{pmatrix} \mathbf{1} & 6 \\ \mathbf{2} & 4 \end{pmatrix} = \begin{pmatrix} \mathbf{19} & 66 \\ 8 & 16 \\ 24 & 80 \end{pmatrix} \tag{B.6}$$

- Matrix multiplication is associative and distributive, but in general not commutative:

$$A(BC) = (AB)C \tag{B.7}$$

$$A(B + C) = AB + AC \tag{B.8}$$

$$AB \neq BA \tag{B.9}$$

## B.2 Matrix Transpose

- The transpose of a matrix A, denoted $A'$ or $A^T$, is obtained by exchanging the rows and columns of A, e.g.,

$$\begin{pmatrix} 7 & 70 \\ 8 & 16 \\ 8 & 80 \end{pmatrix}' = \begin{pmatrix} 7 & 8 & 8 \\ 70 & 16 & 80 \end{pmatrix} \tag{B.10}$$

- If $A + B$ is defined, then

$$(A + B)' = A' + B' \tag{B.11}$$

- If $A$ and $B$ are conformable, then

$$(AB)' = B'A' \tag{B.12}$$

## B.3 Matrix Inverse

- The *identity* matrix $I$ of size $n$ has 1s in all of its diagonal elements but 0s in all off-diagonal elements. It has the property that $AI = A$ and $IA = A$ whenever those products are defined.

- If $A$ is a square matrix and $AB = I$, then $B$ is said to be the *inverse* of $A$, denoted $A^{-1}$. Then $BA = I$ will hold as well.

- If $A$ and $B$ are square, conformable and invertible, then $AB$ is also invertible, and

$$(AB)^{-1} = B^{-1}A^{-1} \tag{B.13}$$

## B.4 Eigenvalues and Eigenvectors

Let $A$ be a square matrix.[1]

---

[1]For nonsquare matrices, the discussion here would generalize to the topic of *singular value decomposition*.

- A scalar $\lambda$ and a nonzero vector $X$ that satisfy

$$AX = \lambda X \qquad\qquad\qquad \text{(B.14)}$$

  are called an *eigenvalue* and *eigenvector* of $A$, respectively.

- If $A$ is symmetric and real, then it is *diagonalizable*, i.e., there exists a matrix $U$ such that

$$U'AU = D \qquad\qquad\qquad \text{(B.15)}$$

  for a diagonal matrix $D$. The elements of $D$ are the eigenvalues of $A$, and the columns of $U$ are the eigenvectors of $A$ (scaled to have length 1). Also, the eigenvectors will be *orthogonal*, meaning the inner product of any pair of them will be 0.

## B.5    Mathematical Complements

### B.5.1    Matrix Derivatives

There is an entire body of formulas for taking derivatives of matrix-valued expressions. One of particular importance to us is for the vector of derivatives

$$\frac{dg(s)}{ds} \qquad\qquad\qquad \text{(B.16)}$$

for a vector $s$ of length $k$. This is the *gradient* of $g(s)$, i.e., the vector

$$(\frac{\partial g(s)}{\partial s_1}, ..., \frac{\partial g(s)}{\partial s_k})' \qquad\qquad\qquad \text{(B.17)}$$

A bit of calculus shows that the gradient can be represented compactly. in some cases, such as

$$\frac{d}{ds}(Ms + w) = M' \qquad\qquad\qquad \text{(B.18)}$$

for a matrix $M$ and vector $w$ that do not depend on $s$. The reader should verify this by looking at the individual $\frac{\partial g(s)}{\partial s_i}$. Note that it makes good intuitive sense, since if $s$ were simply a scalar the above would simply be

$$\frac{d}{ds}(Ms + w) = M \tag{B.19}$$

Another example is the *quadratic form*

$$\frac{d}{ds}s'Hs = 2Hs \tag{B.20}$$

for a symmetric matrix $H$ not depending o $s$, and a vector $s$. Again, it makes good intuitive sense for scalar $s$, where the relation would be

$$\frac{d}{ds}(Hs^2) = 2Hs \tag{B.21}$$

And there is a Chain Rule. For example if $s = Mv + w$, then

$$\frac{\partial}{\partial v}s's = 2M'v \tag{B.22}$$

Now for minimizing in (15.30), use (B.22), with $s = V - Qu$ and $v = u$, to obtain

$$\frac{d}{du}[(V - Qu)'(V - Qu)] = 2(-Q')(V - Qu) \tag{B.23}$$

Setting this to 0, we have

$$Q'Qu = Q'V \tag{B.24}$$

which yields (15.31).

# References and Index

# Bibliography

[1] ARHOEIN, V., GREENLAND, S., AND MCSHANE, B. Scientists rise up against statistical significance, 2019.

[2] BARABÁSI, A.-L., AND ALBERT, R. Emergence of scaling in random networks. *Science 286*, 5439 (1999), 509–512.

[3] BHAT, U. *An Introduction to Queueing Theory: Modeling and Analysis in Applications*. Statistics for Industry and Technology. Birkhäuser Boston, 2015.

[4] BLACKARD, J. A., AND DEAN, D. J. Comparative accuracies of artificial neural networks and discriminant analysis in predicting forest cover types from cartographic variables. *Computers and Electronics in Agriculture 24*, 3 (1999), 131 – 151.

[5] BREIMAN, L. *Probability and stochastic processes: with a view toward applications*. Houghton Mifflin series in statistics. Houghton Mifflin, Boston, 1969.

[6] CHAMBERS, J. *Software for Data Analysis: Programming with R*. Statistics and Computing. Springer New York, 2008.

[7] CHAUSSÉ, P. Computing generalized method of moments and generalized empirical likelihood with R. *Journal of Statistical Software 34*, 11 (2010), 1–35.

[8] CHEN, W. *Statistical Methods in Computer Security*. Statistics: A Series of Textbooks and Monographs. CRC Press, Boca Raton, FL, 2004.

[9] CHRISTENSEN, R. *Log-linear models and logistic regression*, 2nd ed. Springer texts in statistics. Springer, New York, c1997. Earlier ed. published under title: Log-linear models. 1990.

[10] CHRISTENSEN, R., JOHNSON, W., BRANSCUM, A., AND HANSON, T. *Bayesian Ideas and Data Analysis: An Introduction for Scientists and Statisticians.* Taylor & Francis, Abington, UK, 2011.

[11] CLAUSET, A., SHALIZI, C. R., AND NEWMAN, M. E. J. Power-law distributions in empirical data. *SIAM Review 51*, 4 (2009), 661–703.

[12] DHEERU, D., AND KARRA TANISKIDOU, E. UCI machine learning repository, 2017.

[13] ERDÖS, P., AND RÉNYI, A. On random graphs, i. *Publicationes Mathematicae (Debrecen) 6* (1959), 290–297.

[14] FARAWAY, J. *Extending the Linear Model with R: Generalized Linear, Mixed Effects and Nonparametric Regression Models.* Chapman & Hall/CRC Texts in Statistical Science. CRC Press, Boca Raton, FL, 2016.

[15] FARAWAY, J. *faraway: Functions and Datasets for Books by Julian Faraway*, 2016. R package version 1.0.7.

[16] FOX, J., AND WEISBERG, S. *An R Companion to Applied Regression*, second ed. Sage, Thousand Oaks CA, 2011.

[17] FREEDMAN, D., PISANI, R., AND PURVES, R. *Statistics.* W.W. Norton, New York, 1998.

[18] GOLDING, P., AND MCNAMARAH, S. Predicting academic performance in the school of computing. In *35th ASEE/IEEE Frontiers in Education Conference* (2005).

[19] HANSEN, L. P. Large sample properties of generalized method of moments estimators. *Econometrica: Journal of the Econometric Society* (1982), 1029–1054.

[20] HOLLANDER, M., WOLFE, D., AND CHICKEN, E. *Nonparametric Statistical Methods.* Wiley Series in Probability and Statistics. Wiley, Hoboken, NJ, 2013.

[21] HSU, J. *Multiple Comparisons: Theory and Methods.* Guilford School Practitioner. Taylor & Francis, Abington, UK, 1996.

[22] KAGGLE. Heritage health prize, 2013.

[23] KAUFMAN, L., AND ROUSSEEUW, P. *Finding Groups in Data: An Introduction to Cluster Analysis.* Wiley Series in Probability and Statistics. Wiley, Hoboken, NJ, 2009.

[24] KAYE, D. H., AND FREEDMAN, D. A. Reference guide on statistics. *Reference manual on scientific evidence,* (2011), 211–302.

[25] KLENKE, A. *Probability Theory: A Comprehensive Course.* Universitext. Springer London, 2013.

[26] KLETTE, R. *Concise Computer Vision: An Introduction into Theory and Algorithms.* Undergraduate Topics in Computer Science. Springer London, 2014.

[27] KUMAR, A., AND PAUL, A. *Mastering Text Mining with R.* Packt Publishing, Birmingham, UK, 2016.

[28] MATLOFF, N. *The Art of R Programming: A Tour of Statistical Software Design.* No Starch Press, San Francisco, 2011.

[29] MATLOFF, N. *Statistical Regression and Classification: From Linear Models to Machine Learning.* Chapman & Hall/CRC Texts in Statistical Science. CRC Press, Boca Raton, FL, 2017.

[30] MATLOFF, N., AND XIE, Y. *freqparcoord: Novel Methods for Parallel Coordinates,* 2016. R package version 1.0.1.

[31] MCCULLAGH, P. *Generalized Linear Models.* CRC Press, Boca Raton, FL, 2018.

[32] MILDENBERGER, T., ROZENHOLC, Y., AND ZASADA, D. *histogram: Construction of Regular and Irregular Histograms with Different Options for Automatic Choice of Bins,* 2016. R package version 0.0-24.

[33] MITZENMACHER, M., AND UPFAL, E. *Probability and Computing: Randomized Algorithms and Probabilistic Analysis.* Cambridge University Press, Cambridge,UK, 2005.

[34] MURRAY, J. F., HUGHES, G. F., AND KREUTZ-DELGADO, K. Machine learning methods for predicting failures in hard drives: A multiple-instance application. *J. Mach. Learn. Res. 6* (Dec. 2005), 783–816.

[35] MURRELL, P. *R Graphics, Third Edition.* Chapman & Hall/CRC The R Series. CRC Press, Boca Raton, FL, 2018.

[36] ORIGINAL BY GARETH AMBLER, AND MODIFIED BY AXEL BENNER. *mfp: Multivariable Fractional Polynomials,* 2015. R package version 1.5.2.

[37] ROSENHOUSE, J. *The Monty Hall Problem: The Remarkable Story of Math's Most Contentious Brain Teaser.* Oxford University Press, Oxford, UK, 2009.

[38] SARKAR, D. *Lattice: Multivariate Data Visualization with R.* Use R! Springer New York, 2008.

[39] SCOTT, D. *Multivariate Density Estimation: Theory, Practice, and Visualization.* Wiley Series in Probability and Statistics. Wiley, Hoboken, NJ, 2015.

[40] SHANKER, M. S. Using neural networks to predict the onset of diabetes mellitus. *Journal of Chemical Information and Computer Sciences 36*, 1 (1996), 35–41.

[41] WASSERSTEIN, R. L., AND LAZAR, N. A. The asa's statement on p-values: Context, process, and purpose. *The American Statistician 70*, 2 (2016), 129–133.

[42] WATTS, D. J., AND STROGATZ, S. H. Collective dynamics of 'small-world'networks. *nature 393*, 6684 (1998), 440.

[43] WICKHAM, H. *Advanced R.* Chapman & Hall/CRC The R Series. CRC Press, Boca Raton, FL, 2015.

[44] WICKHAM, H. *ggplot2: Elegant Graphics for Data Analysis.* Use R! Springer International Publishing, 2016.

[45] XIE, Y., LI, X., NGAI, E. W. T., AND YING, W. Customer churn prediction using improved balanced random forests. *Expert Syst. Appl. 36* (2009), 5445–5449.

# Index

Printed in the United States
by Baker & Taylor Publisher Services

Printed in the United States
by Baker & Taylor Publisher Services